The Historical Christ

Conservation and Restoration of Insect Population

Conservation and Restoration of Insect Population

Edited by:

Laichattiwar Mukesh Anandrao

Delve Publishing

www.delvepublishing.com

Conservation and Restoration of Insect Population

Laichattiwar Mukesh Anandrao

Delve Publishing
224 Shoreacres Road
Burlington, ON L7L 2H2
Canada
www.delvepublishing.com
Email: orders@arclereducation.com

ISBN: 978-1-77407-759-7 (Hardcover)

ABOUT THE EDITOR

Dr. Mukesh Laichattiwar, Ph. D (Institute of Agricultural Sciences, Banaras Hindu University, Varanasi, U.P.) is presently working as Assistant Professor at College of Agriculture, Parul University, Vadodara. Dr. Laichattiwar worked as Assistant Professor at college of Agriculture, Naigaon (Bz.), Nanded M.S. (2018-2019). He did his B.Sc (Ag.) From College of Agriculture, Ambajogai (Vasantrao Naik Marathwada Krishi Vidyapeeth, Parbhani, M.S.) and M.Sc (Ag.) from Institute of Agricultural Sciences, Banaras Hindu University,Varanasi, U.P., He was awarded UGC fellowship for his Ph. D research. He qualified ICAR-NET in 2013. Dr. Laichattiwar is a recipient of Best Poster presentation award in 2014. He Published 18 research article, 7 book chapters/proceedings, 5 popular articles and 15 abstract.

TABLE OF CONTENTS

LIST OF FIGURES

wasps have been found to aggressively compete with honeybees for the honeydew that is secreted in New Zealand's South Island black beech forests by the scale insect Ultracoelostomabrittini

Figure 9.8: Above shows some of the insects that are disappearing at a high rate. Pollution is accounting for the majority of this decline, mainly due to global warming which is either driving the insects away, or reducing their life span. By 2019, the total decline was 41% as seen above. Caddisflies had shown the greatest decline

Figure 9.9: As you can see above, the use of pesticides, fertilizers and also agricultural intensification accounts for the major threat on insect populations, being the biggest challenge in the conservation of insects

Figure 10.1: Ornithoptera meridionalis, the southern tailed birdwing, is the smallest species of the genus Ornithoptera. It is known from a handful of localities in southeast Papua, New Guinea and several localities along the south coast of Irian Jaya. Its bigger counterpart, Queen Alexandra's Birdwing Butterfly is also protected in under the Convention on International Trade in Endangered Species (CITES)

Figure 10.2: Justifying insect conservation and acting on this justification requires a philosophical view based on value systems on nature. Valuing the human relationship with nature results in benefits accruing in terms of physical resources such as food and also in terms of well-being on the mental, emotional and even spiritual levels.

Figure 10.3: The abundance of the Large Blue butterfly in the United Kingdom, preceding extinction in 1978 and following its reintroduction from Sweden to carefully managed sites in the 1980s. No precise data are available for numbers prior to 1972, but anecdotal descriptions indicate a continual decline, with a similar number of butterflies to today having last occurred in the early 1950s.

Figure 10.4: the table above shows the number of state insect pollinator policies passed and approved by Legislatures and signed into law by Governors by year

Figure 10.5: Insect conservation is transferred from policy to practice is carried out by different organizational communities. Land cores, like protected areas and the inner zones of reserves, are critical for many insect species. These areas often require some management to recreate natural conditions. Outside of these land preserves, other approaches to landscape design and management can be used as insect conservation action. These approaches include agro-ecology, organic agriculture, the land sparing-land sharing techniques, agro-environment schemes, restoration, urban improvement, and removal of invasive non -indigenous plants

PREFACE

Insects are one of the smallest species of animals but perhaps the most exploited kind of them all. Insects dominate the ecosystem and provide a hub that we humans and other animals enjoy to fill up the circle of life. These three-segmented animals dominate the entire world from pole to pole covering the immense cold to the super-hot regions where even man has failed to show his persistence to nature. Their immense numbers covering up to 90% of the animal population play key roles in the natural world where humans are slowly tearing apart at an alarming rate.

Food is the fuel on which animals propel on. Insects feed up a bigger percentage of the ecosystem providing food to amphibians, birds, reptiles, some fish and even mammals such as human beings. They further help decay dead matter clearing the world of air pollution and providing key nutrients to plants, giving back what nature has borne unto them. Insects also pollinate plants that later bear fruits and food and some further destroy crop pests that lag food production behind.

It can't be argued that insects are highly endangered yet we also can't imagine a world without insects. The environment itself brings around hazards either natural or manmade. Forest fires such as the sporadic Australian fires, flooding, global warming, deforestation and human encroachment to their habitats are some of these hazards. What comes out is that humans contribute a lot in breaking the circle of life. Our technological aspects are perhaps important to our lucrative nature but insects could cover up on the side effects of these hazardous aspects.

One thing clear is that insects should be conserved, but how do we go about it? One of the ways is through integrated control. This is whereby different methods are used. One of the methods involves, controlled use of pesticides, that is, avoiding watercourses and hedges and more inclusive education. Reducing greenhouse gas emissions levels down the rate of global warming bringing normalcy on weather patterns and further restoring insect cycle. Government policies and legislation step in limiting and preventing human encroachment and deforestation on riparian land and natural habitat. Parliament and law enforcement agencies control such activities by making and implementing laws that cover natural habitats and punish those who go against the same.

With a world that focuses on research and new developments, bodies and organizations that cover up such should refrain from introducing invasive species that may hinder the cycle of important insect species. As much as new insect species developed from laboratories may be of importance it doesn't prevent the fact that a big win may always bring a massive loss. However, we may popularize insects through the use of various iconic species.

Insects and humans or rather the massive population intertwine on the same pillar. One side of a burning coin will always make the other side hot. As much as it is about land managers, government stakeholders and the podium bearers, it is more of us human beings taking up the role in insect conservancy.

1

Introducing Insect Conservation

CONTENTS

1.1 INTRODUCTION

Figure 1.1: Invasive aphids exploring their new habitat.

Source: https://www.nwf.org/Educational-Resources/Wildlife-Guide/Threats-to-Wildlife/Invasive-Species

Conservation of insects isn't a completely new concept. It's been in existence for more than a century, but to some extent, it has been and still to some extent, just like the insects it's attempting to save, not a very popular activity. But now insects are gradually being recognized, and the significance of protecting insects is increasingly accepted by those in power. Individuals regularly respond to the notion of "insect conservation" by asking two fundamental questions, "What is it?" and "How is it beneficial?" In this chapter, the objective is trying to respond to these two concerns.

Of course, there are many insects around, after all, you have been seeing hundreds of insects in the field and the forest in your area, which is a characteristic non-entomological reaction to the concept of "insect conservation". In general, the variety of life on earth is an important reason for conserving insects, reducing diversity also means reducing the joy of everyday life. The economic impact of insects is equally enormous, they help in pollination of both cash and subsistence crops used by humans. The honey industry, for instance, relies solely on bees and is a huge industry that provides jobs for 1000s of persons and generates billions of dollars in profits each year (Cheesman, & Key, 2007).

1.2 KNOWING INSECTS' BIODIVERSITY

Biodiversity translates to all forms of life on the planet. It comprises the diversity of species and ecologies in the world (communities and interconnections of species), as well as genetic variance. Insects are the biggest and most varied biota on the planet. There are about 30 orders, and the amount of described species reaches nearly 1 million. Nevertheless, some researchers believe that the real number may surpass 20 million. These amazing creatures account for around 75% of all existing animal species described and are extremely abundant in number, but they have been neglected. Insects have been found in water and atmosphere in almost all environments and on all continents, covering Antarctica as well (Collins, 2012).

Just like amphibians, the richness of insect species is highest in tropical regions. Similarly, in the neotropical region, the richness of insect species is high, although the real number of species defined at this time is still uncertain. Ecuador is also a member of 17 major biodiversity countries and can have a large number of insect groups. But what exactly are arthropods? Arthropods are invertebrates having sectioned bodies, joints and chitinous exoskeletons. Apart from insects, arthropods comprise of spiders, millipedes and mites.

Insects are the biggest group of arthropods; however, they can be differentiated from other types of arthropods through certain features. They consist of 3 body parts (head, chest and abdomen), 3 sets of legs and two antennae. Among the gifts, most insects have which other arthropods don't is the capacity to fly. In matters of reproduction, mammals ,as well as other vertebrates, are more understandable compared to insects. Most insects are oviparous (lay eggs outside the female body), but some insects are viviparous (no egg-laying involved) and can produce live offspring. Some insects furthermore reproduce through parthenogenesis (no males are present).

Surprisingly, careful parental care occurs in certain insect species. Such a massive and varied group shows that their environments are very complex. Therefore, insects play many significant roles in these environments. Most are herbivores, whereas others feed on predators, other insects and arthropods, and even certain vertebrates. Many species are parasites on host vertebrates, while a few are parasitoids meaning females produce eggs on these arthropods feeding on them and these eggs hatch inside their hosts. Their function as pollinators can't be underrated. Many taxon units act as scavengers and even decomposers (Collins, 2012).

Views on insects vary widely. They are avoided and dreaded in some cultures and esteemed in others. Indeed, some insects contend with people for food, especially in agricultural environments, plus many insects are vectors of severe pathogens. Nevertheless, we must reflect on their general value in any particular ecosystem to which they fit. For the ecosystem to function properly, proper conservation and comprehension of this particular group are critical.

Insects are essential due to their wide variety, ecological function, and impact on agriculture, health, and natural properties. This chapter discusses the impact of insects in the environment and explains how insects have become central in many scientific advances. Insects make up the biological foundation of every land ecosystem. They circulate nutrients, scatter seeds, pollinate plants and preserve soil texture and fertility, among other functions (Collins, 2012).

1.3 DEVELOPING APPRECIATION FOR INSECTS

Scientists approximate that there are roughly 5 to 30 million types of insects worldwide, most of which have not been scientifically collected or named! Generally, the most famous insects are those which cause illness such as mosquitoes or contend for human agricultural resources, however, these insects only account for a small part of the globe's insect population. In fact, many insects are helpful to people and the ecosystem and are not dangerous. Without insects, the ecosystem will not work, but few people consider how the diversity of these many creatures affect our existence.

Among the most significant services offered by insects is pollination. Although some plants self-pollinate or are wind-pollinated, numerous flowering plants depend on insects to transfer pollen to different flowers, thereby promoting fertilization. Insects tour flowers to gather nectar, however many insects have specified anatomical structures to carry the pollen they inescapably encounter. Bees usually collect pollen in a "basket" developed by bristles on their rear legs or abdomen. Speckled beetles absorb pollen with many bristles, and the developed scales resemble hair. Some plants, like orchids, have further co-evolved with insects, plus are only pollinated through a specific insect.

After observing a garden species in Madagascar, scientist Charles Darwin foretold that a moth with 28-cm (11-inch) long proboscis would be found there, equal to the length of the orchid's corolla tube. Since then, scientists have discovered this moth, which is believed to be able to pollinate that specific orchid. For humans, bees are the most important pollinators. Without bees, many plants we depend on cannot reproduce and produce most of the fruits we consume, like apples and strawberries.

In fact, insects also make an important aspect of food chains and an instant food source for several animals, humans included. In the U.S., insects are mainly consumed as a novelty, while in other communities, insects like beetles, termites, earthworms and honeybees are usually made and consumed as food. However, the most important food item from insects is honey, which is produced during all pollination processes by nectar collected by bees from flowers. Another significant advantage of bugs is their function in breaking down organic material. Museum staff has used this fact to clean the bones of mammals using developed carpet beetle colonies.

Without insects, the earth will be overwhelmed by garbage-decomposing leaves as all kinds of waste are consumed by insects. Microorganisms are the ultimate decomposers, however, when larger creatures first prepare the decomposed plant material to a big extent by consuming leaves and making fecal particles, the efficiency will increase. Since insect corpses are mainly made of nitrogen, their demise will return this ingredient into the earth.

After several years of cicada decomposing in the eastern U.S., large amounts of nitrogen got recycled back to the earth in 2004. Nature's balance relies on the functions of parasites or predators, many of which are bugs. A number of insects, such as caterpillar hunter beetles, mantises and pirate bugs, regulate the population of herbivorous insects. Similarly, important are parasites. The most ideal example is the braconid wasp that hatches eggs on the tomato-hornworm.

1.4 ETHICS OF INSECT CONSERVATION

Entomologists have continually rated insects highly, among all creatures (after all, compared with other scientists, insects are closer to them). But until lately, most scientists would still say that despite E.O Wilson 's 30-year conservation effort for the 'little things running the world' which calls for a focus on insects, the general public still has no such idea. In a flash, the public was shocked by the reported downward trend in the number of insects worldwide. Taking advantage of this emergence of public attention and social interest in discoveries of insects will provide immense new opportunities to enhance insect conservation more widely, however, this will not continue without a couple of challenges- a few of which may restructure the field as it's known today (Samways, 2019).

Throughout the world, many insect museums are packed with a large number of "wet collection" insects of thousands of specimens, though these collections are hardly used as resources. The result of the formal establishment of the 3Rs framework may be increased reuse of archived objects. The creation of some or each of these insect conservation areas will undoubtedly require new approaches in ethical standards. Basically, researchers have proposed five pragmatic methods that enable other researchers to enhance their own principled sampling methods and reduce the damage to insects. Some safety measures to be considered when handling insects include:

- Perform an efficacy analysis to decide whether the overall sampling can be reduced;
- If you must use existing sampling techniques, consider changing the deployment agreement to diminish by-catch;
- Adjust to a more precise trapping technique to prevent by-catch;
- Provide by-catch for future research use; and
- Reduce the pain of insects during the collection stage

Ultimately, entomologists must work out the way forward. The researchers only encourage insect experts to observe the societal and philosophical significance of ethics in insect dynamics and debate an appropriate context and actions which must be considered to lessen harm and reduce conservation risks.

In this understanding, one of the emerging issues being seen is that people are increasingly concerned about the ethical handling of insects in research activities. Questions like "Why must we kill endangered insects for research when the goal is protecting them?" "Because of modern research, why should insects be brutally tortured?", plus "Why must we kill such a high number of non-target insects?" With the public being under growing pressure to safeguard insects, it's becoming an increasingly acute issue. This shall inevitably affect how scientists conduct investigations on insects and publish outcomes (Samways, 2019).

In a world of reduced insect populations, it's worth considering the long-term feasibility of trapping methods, which are one or two sequences of magnitude greater than the non-target insects actually captured by the target. When this method is essential, researchers believe that when the captured organisms are used to the maximum, it is easier to justify large-scale trapping, for example by creating a “non-target” component register of bulk samples that can be utilized in the future by different researchers. In a broad sense, there’s a need to follow the codes of conduct like not indiscriminately gathering rare insects, since this could cause population decrease or extinction.

1.5 UNDERSTANDING THE PAST AND PRESENT DISTRIBUTION OF INSECT BIODIVERSITY

Figure 1.2: Dragonflies can be used to measure the health of aquatic ecosystems.

Source: https://theconversation.com/meet-the-insects-that-are-defying-the-plunge-in-biodiversity-new-findings-131846

Protecting biodiversity is one of the most difficult challenges in society, and this can only be solved through multidisciplinary efforts. Robust science will be the foundation of sustainable policy, but in scientists’ comprehension of the complicated ecological dynamics that operate on various spatial and temporal balances from subcellular to terrestrial, from seasonal to era, there are still many scientific uncertainties. This uncertainty is beyond the scope of conventional disciplinary thinking and requires the joint insights of ecologists, evolutionary biologists and conservation naturalists and taxonomists among other scholars (Samways, 2019).

Some key questions to ask are: How shall species and communities react to climate change? What impact does the loss of biodiversity have on ecosystem functions and the provision of key environmental systems? Which species are highly likely to become extinct, and how to restore damaged ecosystems? Solving these and other persistent issues will certainly need a flourishing and multidisciplinary community of biodiversity researchers.

1.5.1 Lessons from History and the Revival of Natural History

Without understanding the loss and increase of species during a geological time, any appraisal of the standing of biodiversity is inadequate. The realm of paleoecology can be traced back to the past to offer this view and provide us with views on the current dynamics of biodiversity within the global change era. The plenary session of the United Nation's World Heritage Dinosaur Park, Alberta highlighted the lessons learned from environmental changes that happened around 70 million years back, opening a window to the early biodiversity of the earth. The meeting also mentioned that 125 years following the discovering of Albertosaurus, new species of animals including insects continue to be found.

Another symposium on paleo-biodiversity additionally explored the roots of contemporary biodiversity by introducing species combinations and dynamics around different taxa and time frames, and also discussed in detail the method of rebuilding history from fossil records. The realms of ecology and evolution have been born from the research of natural history; meaning, the species and their circulation and interrelationships were carefully described. These root causes are regularly forgotten in the realm of advanced technology today; however, we still rely on these fundamental foundations.

1.5.2 The Future of Biodiversity

Ecologists are becoming more and more aware that the understanding of ecological evolution patterns and processes is not sufficient in itself to address fears about the loss of biodiversity. Therefore, some environmental conference discussions are dedicated to the application of ecology and evolutionary science in solving environmental problems. Scientists ,for instance, have put the theme of ecological impact at the forefront of their presentation, addressing the effect of oil sands industry activities on the Athabasca River ecosystem in Alberta.

The researchers believe that decades of insect population ecology research has provided a solid foundation for solving application problems, however when condemning the "biodiversity ignorance" and the things we do, they emphasize the incompleteness of our understanding of ecology (For example, the driving force behind insect population decline). Such a lack of understanding is inefficient in influencing policies. They also raise concerns about the difficulty of gaining long-term support for insect biodiversity monitoring for their respective jurisdictions. When solving management problems, it is also necessary to merge long-term monitoring with original datasets and inquiry.

1.6 IDENTIFYING HARMFUL FACTORS INFLUENCING INSECTS' POPULATIONS

1.6.1 Habitat Loss

Scientists have cautioned that environmental degradation and losses are destroying the ecosystems that many insects depend on. Among the known fundamental causes is the loss of habitat. This situation is especially due to the decrease in the number of insects and the extinction caused by landscape division and wide grounds of single crops, resulting in landscape degradation and ultimately loss of their native habitat. Its impact on insect populations is complex and interrelates with other influences. Climate change is related to more extreme weather patterns and more extreme and frequent fires that reduce insect populaces. It also alters the prevalence of pests, making their management more challenging (Kumar, Yogi, &Jagdish, 2013).

In addition, the loss of habitat means that insects can't travel on the terrain as easily as they used to locate the conditions which best suit them. Their best habitats are getting more apart and reduced. However, the future is not desperate at all. Strategies are getting developed all over the world, and if expanded, they shall benefit insects worldwide. Whether in agriculture or forestry or in cities, certain species can survive well in the environment of human alterations. Others can survive in some agricultural ecosystems and even urban parks. However, many need a special environment or a specific host species to survive.

These particular insects are those lost at a very alarming rate, particularly in tropical forests where deforestation is rampant. Their residential space is greatly reduced, decreasing their chances of survival. When the shrinking level reaches a critical level, they have no choice. On the contrary, certain genetic modifications allow certain insects to adjust to the shifting human environment. For example, the tiny ferret moth responds less and less to artificial light, thereby increasing its opportunities for survival in an urban setting. Others can benefit greatly from certain artificial settings. In the example of artificial ponds, this is commonly seen. Studies have established that these offer many more prospects for survival, since more options are accessible, particularly when ordinary ponds are undergoing drought stress (Kumar, Yogi, &Jagdish, 2013).

Scientists have suggested a road plan to solve most of the problems facing insects. These strategies not only provide a direction for the enduring survival of insects but also ensure that insect populaces continue to offer ecosystem solutions beneficial to people. These measures include pollination of crops, the use of natural predators and parasite insects to control pests and maintain healthy soil. However, recently, more detailed policies have arisen. They focus on particular ecosystems, whether they are forests, freshwater, grasslands, cities or caves. In brief, the combination of different research activities across the world and effective implementation demonstrates the progressive way forward.

These strategies comprise of more investment going into the future, but they can help salvage insect populations in their respective habitats. Among the available strategies, a network of functional natural vegetation corridors can be implemented between crops and plantations to allow insects to move through the landscape. It can also be beneficial to plant special vegetation between the rows of crops and surrounding the edge of fields, including careful planting along the road (Reddy, 2017).

1.6.2 Habitat Fragmentation

Habitat fragmentation happens when formerly linked habitats are split due to human development (such as city areas, farmland, logging yards, and roads). This only allows for smaller, isolated regions that insects can effectively inhabit "islands". The marginal areas of these islands are exposed to stressors like human interference, alien species and pollution. The debris interferes with the movement, foraging and stalking habits of predators that keep insect populations in check. People are beginning to notice that isolated parks encircled by human developments aren't enough to protect wildlife. It is the natural habitats and wildlife corridors that need to be interconnected for species to thrive.

The fragmentation of habitats and the increase in agriculture are some reasons behind the decrease of insect generalists. Based on a study presented in the Scientific journal for Biological Conservation, the genomic diversity between the butterfly species examined is also projected to drop suddenly in the near future, and as a consequence, these insects shall become more exposed to environmental variations. In some areas, the number of insects continues to decrease. In recent decades, the number of insects has fallen sharply by 75%. Until presently, scholars still think that this has only been about specialist insects, that is, insects that rely on particular habitats are more vulnerable to extinction (Reddy, 2017).

In the future, even the supposed "ubiquitous species" may face huge threats. In a research article, scientists mentioned that species having low habitat needs depend mostly on exchanges between diverse populations. The study clearly shows that widely distributed species have a more varied genome pool. Once lost, these animals maintain this opportunity for genetic diversity and will not be able to acclimatize to varying environmental circumstances going into the future.

Insect scholars from Munich mentioned that "the underlying causes of species decline have changed over time. Originally, it was mainly those insects that specialized in specific ecosystems, for example, Parnassiusapollo butterflies, which face a risk with an increased deterioration of habitats and the collapse of the whole habitat network. The threats to "oversized" species such as "Coenonymphaarcania" remain widespread.

In terms of actual nature conservation, research indicates that a small number of isolated protected areas won't be sufficient anymore in the future, and these protected areas will benefit specific species with fundamental genetic structures, relying on most of the exchanges between native populations. Environmentalists predict that in the moderate and long term, this situation will disappear, which will lead to an increased reduction in many insect species, this will have a huge impact on the whole food web and insect ecosystems (Reddy, 2017).

Loss of habitat and its degradation are probably being the most significant dangers to insect biodiversity. Universally, up to 50 percent of the endemic plant species and vertebrates are limited to about 36 biodiversity hotspots that cover roughly 2.5 percent of the Globe's surface. It can be said that these hotspots may contain similar proportions of prevalent insect species. The latest model shows that the economic pressure of agriculture on land will further diminish the presently constrained natural intact vegetation in one-third of the globe's hotspots by 50% come 2050. Processes related to deforestation, agricultural growth, and modernization are direct drivers of the disappearance of natural or partly-natural habitats and their respective insect groupings (Reddy, 2017).

1.6.3 Disturbance in Habitat Quality

Natural habitats are physical, chemical and biological systems that support organisms like insects. In short, habitat is where these creatures live. When natural or human activities alter these places, habitats are lost and degraded, therefore reducing the population of species living there. For instance, when a forest is expunged and replaced with agricultural land or urban spaces, multiple insect species may be eradicated. Although natural factors like landslides and earthquakes may change the landscape, they usually happen in remote areas where healthy ecosystems can recover. Human-instigated habitat loss, meanwhile, is altering ecosystems on a universal scale, typically causing devastation that is permanent, at least on human time scales (Saunders, Janes, & O'Hanlon, 2020).

From a concerned perspective, the loss of habitat is not just a worry for insect species that are part of the planet. People are also part of the earth's life cycle, therefore, the general function of the natural system on which we depend. Normally functioning natural systems produce the air that we breathe, decompose our waste, offer food, purify drinking water and eventually provide all the resources we need to live. Every species plays a major function in its environment. Habitat disturbance and degradation are among the key threats to the globe's endangered insects and plants, plus this threat is happening at an increasing rate.

Despite the sobering situation, habitat degradation is not constantly an inevitable result of human activities. There are many ways to reduce our influence on natural insect ecosystems, for instance by using the "smart growth" principle in urban planning. However, significant changes and efforts still are needed to resolve this crucial issue (Saunders, Janes, & O'Hanlon, 2020).

1.7 HOW DOES HABITAT DEGRADATION AFFECT ECOSYSTEMS?

Habitat can be damaged in many ways. When the habitat changes physically, like when a wetland becomes filled or the coastline is substituted by a seawall, direct destruction occurs. In addition, pollution can make the environment too toxic to all or some species, resulting in indirect damage. Invasive species can change the structure and chemical properties of ecosystems and compete with native species. Estuaries and wetlands make up the most dynamic ecosystems around the globe, providing many services that directly benefit people; for instance, water sieving and storage, and safety from stormsand erosion. Apart from that, they allow for the multiplication of different insect species.

The habitats presented by these ecosystems is diverse. In the intertidal zone, all kinds of insects inhabit the submerged eelgrass beds and the salt marshes of the estuary. Wetlands also contain a variety of endangered insects. Together with many microorganisms and plant species, they form a complex food network. Damage or removal of a component in the food web will affect the workings of the entire system. Regrettably, in some cases, people ignore the benefits of estuaries and wetlands, they simply treat it as a "waste" space and can use dikes, drainage and fill it up for agriculture, modern development or other "beneficial" goals while disregarding the insect biodiversity they support. When some kinds of plants or insects are eradicated from these effects, then other species also that rely on them may disappear or, where possible, migrate to another area.

1.7.1 Loss of Critical Resources

All species (including insects) are valued as unique blends of evolutionary processes that have innate value, therefore, they need care and protection. However, as George Orwell said in The Animal Farm, "All creatures are equal, but others are more equal compared to the rest." Invertebrates are involved in conservation efforts universally but the insects somehow remain neglected. There's no reason for insects to receive less attention compared to birds or mammals. Nevertheless, the significance of the decline in insect populations and the subsequent extinction far exceeds the loss of species as well as their intrinsic value.

Each species represents lost individuals, biomass and environment function, and thus cannot be used for other organisms. Every species contributes a specific aspect to the complex life balance, and this tapestry will change with time and space. Every species signifies an unrepeatable aspect of life history. Consequently, each species also interrelates with other species plus their surroundings uniquely, creating a complex network that maintains the survival of other species. Scientists have shown that in the coming 30 years, flying insects will lose roughly 75 percent of their existing biomass. This negative trend has been observed mostly in the natural reserves of Germany. These findings present a possible threat and inspired an intense debate around the insect crisis.

In addition, in other parts of Germany, a large number of arthropods and biomass decline have been documented. Similar tendencies have been mentioned in other sections of Europe. Recently, the abundance of butterflies, moths and beetles in the UK has also dropped significantly. Negative trends aren't limited to Europe, however, also happen in other sections of the globe. A universal meta-analysis of the number of insects reveals that in the three kinds of units evaluated, the abundance of insects has declined by about 45 percent.

However, the particular trend and intensity of the decline and final rise are not universal and will vary depending on taxa and regions. The decline in the number of insects and biomass constantly occurs before the extinction of the species, because this is a continuous process, not a binary process. Although heavily reliant on the ecological function of species, the digital loss of abundance and the extension of biomass reflects the damage to ecological functions and the provision of ecosystem services. For instance, biomass is a measure of the energy flowing through the trophic level represented by insects. Consequently, reduced numbers and biomass impacts ecosystem functionality and flexibility, food chain structure, as well as species relations, like plant-pollinators, population persistence, plus many ecosystem services.

The study emphasizes the decline in the value of abundance and biomasses the landscape stage, but also tells us that the decline is not limited to rare or endangered insect species, but also exists in more abundant species. Although insect conservation often targets attractive, rare or endangered species, the spatiotemporal tendencies of regular and prevalent species are frequently overlooked. The decline in the number of regular and widespread insect species has more intensely affected ecosystem functions. Therefore, the protection of ecosystem functions might not receive attention, which highlights the need to monitor and conserve insects, not just rare and endangered species.

1.7.2 Impact of Invasive Alien Species

Figure 1.3: An invasive bee species can negatively affect a region's pollination trends.

Source: http://bsed.org.in/impacts-of-invasive-alien-species-on-honey-bees/

Invasive alien species or (IAS) are artificially introduced species whose locations outside the natural geographic range have obvious environmental, ecological or socioeconomic impacts. The impact may be instant (for example, through predation, competition and disease vectors) or indirect (for example, through nutritional cascades, herbivore or parasitic extinction). Introduced species may eventually cause the local decline of resident insects, while those displaying a narrow geographic distribution or specialized eating habits are the most vulnerable to decimation.

Direct competition from non-native species may drive the local population to extinction. The extent of ecological overlap with invasive ladybugs is the main predictor of the native extinction of endemic ladybug wildlife in the UK. Invasive ants (such as the Argentine ants) are possibly the best illustrations of IAS challenging local insect populations. Because of their massive numbers and accepted predatory behavior, most alien ant species are the main risk to native insects. Invasive amphibians kill aquatic insects significantly more than their natural amphibian predators, reducing the diversity of native insects and replacing natural amphibians.

The haplotype of invasive alien plants reduces the amount and/or value of food and leads to a reduction in the necessary resources of many insects. In addition, invasive alien plants change the composition of the matrix and adversely affect the relationship between the insect host and parasite. Invasive alien plants might also supply ecological evolution traps for local insect species. Once the invader defeats the local host and loses its ability to compete, it may become a host that causes larval dysplasia or greater larval mortality, resulting in a reduction in the insect population. Invasive pathogens may also cause natural insects to become extinct. The European fungal pathogen Nosemabombi strain is thought to cause an extensive decline of the North American hornet.

In addition, due to direct competition or the introduction of pathogens that insects cannot defend against, the introduced bee *Bombus terrestris* L. caused the native Patagonian bee *B. dahlbomii* GuérinMéneville to disappear within most of its home. Moreover, specialization has caused many insects to become co-dependent animals, so it is easy to become extinct together. Among them, countless insect pedigrees and vertebrates have diversified, whether it is parasites, endemic symbiotic wildlife or symbiotic wildlife. Around 5000 lice (Phthiraptera) species have currently been defined, most of which rely on an avian host. The types of insects that commonly depend on plants are also very abundant, as much as 211,000 insects have been found to be gall-inducing.

Although there are few examples of co-extinction, insects that depend on symbiosis are in danger of extinction due to their special ecology. Models show that extinction events should be more common than current records indicate (particularly in plant-reliant beetles and fowl lice). This may be due to poorly documented extinction events, or unrecognized network resilience due to symbiotic insects being able to utilize multiple species than earlier envisaged. For inter-dependent insects, the trophic cascade may be especially relevant. Host species can be lost because of habitat loss, like has been revealed in Lepidoptera-host plant structures. One historic example of indirect impacts of invasive species has been the co-extinction event of Christmas Island flea, caused by loss of Christmas Island rat after the introduction of the black rat.

1.7.3 Climate Change

Climate change presents a risk to insects and environments on which they exist, whether on land, freshwater or underground. The complexity of universal climate change is far beyond the scope of universal temperature surge. It further causes a variety of ecological responses to ecological changes, comprising changes in species distribution, phenological migration, novel interactions between formerly isolated species, extinction, plus other impulsive cascades at various levels of the environmental structure

(Bjorkman & Niemela, 2015).

Variations in species phenology, circulation, decrease in body size, reduction in combined structure, and dysregulation of species-specific relations are all related to climate change. Case in point, some British butterflies appear earlier than formerly recorded, plus in some instances, prior to their nectar plants blooming. Additionally, changes in the diversity of functional nourishment groups may be related to changes in nutritional relations in the food chain. Because of the synergistic negative impact on the overall freshwater ecosystem, aquatic insects are unreasonably impacted by climate change, plus these insects have reduced the ability to spread. And they face transmission barriers, especially at high altitudes.

Biological indicators need to be developed and implemented, and dragonflies are becoming taxonomic advocates of aquatic ecosystems. Researchers call dragonflies "climate canaries". The sensitivity of dragonfly species collections to climate variables is three times that of family-level invertebrate collections. Although there's proof that improved water quality offsets the recent climate debt of aquatic invertebrates, this continued relief is unlikely to reverse or prevent the decline of water insect species (Bjorkman & Niemela, 2015).

After analyzing 50 years of data from the UK, the study warned that insects "have nowhere to hide" because woodlands are not safe havens, and the changing time of emergence damages nature and agriculture. The research found that the shadows of woodlands are supposed to safeguard species from the effects of warming, while the effects of climate change are the same as those of open grasslands. This study surveyed the first spring flight records of butterflies, aphids and moths and the initial eggs of fowls between the years 1965 to 2012. As the average temperature rises, aphids now appear a month before time, while birds lay their eggs up to a week earlier.

Scientists say this may mean that animals and prey are "out of sync" and may have serious consequences for the ecosystem. Scientists are increasingly concerned about the sharp decline in insect populations, which is the foundation of nature. It was noted that in February 2020, these patterns led to "catastrophic decline of natural insect ecosystems", and by March 2020, there was further evidence that widespread pollinating insects in Britain have been lost in recent decades.

Other researches from Puerto Rico to Germany have shown that this number has declined over the past 25 up to 35 years. Yet another shows that over the past 130 years, butterfly numbers in the Netherlands have fallen by around 84%. Under global warming, it may be hoped that the woodland has a certain protective effect on insects and can resist changes. But British scientists have not seen it. This is the biggest surprise and is also disturbing. If you are an insect, then indeed, there is no place to evade the impacts of global warming (Ziska& Dukes, 2014).

1.7.4 Aquatic and Terrestrial Systems Contamination

Due to the heavy use of pesticides and inappropriate risk assessment principles, pesticides are the main driving force behind the decline in insect numbers. Pesticides affect insect populations through direct toxicity and secondary lethal effects (mostly pesticides), and indirectly (mostly herbicides) through habitat changes. Due to long-term exposure and biomagnification in the food chain, bioaccumulation poses serious additional threats to insect populations. These threats may have undetectable harmful impacts on insect functioning and behavior (Collins, 2012). Most fertilizers (such as organic and mineral ones) commonly used in agriculture may indirectly affect insect populations by affecting the composition or plant resources quality, affecting habitat structure, or causing ground acidification and eutrophication. For some herbivorous insects in agricultural ecosystems, the large use of chemical fertilizers may have a positive effect, but it has a negative impact on most insects. In addition, the use of antihelmintic compounds (example, ivermectin) in livestock systems have a negative effect on the abundance as well as the diversity of insects related to fecal decomposition.

Industrial pollution (such as air pollution, compounds from factories and mining activities, and heavy metals) furthermore leads to a decline in insect populations. Similar to pesticides, it has a fatal negative impact on individuals, as well as biological amplification across the food chain, further increasing the risk to insect populations. Various economically significant insect species (like pollinators and natural pest enemies) can be threatened through longstanding exposure to pollutants (like heavy metals), however, the community-wide impacts remain unclear. Freshwater invertebrates (such as various insect taxa) are extremely impacted by pollution, with more than 41 percent of species listed on the endangered insect (IUCN Red List) by water pollution (Collins, 2012). Industrial wastewater, sewage, agrarian and urban runoff, and increased sedimentation have all reduced the quality of freshwater habitats. Light and sound pollution are becoming more and more common on a global scale, so a better comprehension of these new effects is crucial. Night insects are particularly susceptible to changes in the natural light-dark cycle. Light pollution can interfere with insects that utilize natural light (coming from the moon and stars) as navigation directions, and interfere with the spread of insects that use bioluminescent signals (such as fireflies). It destabilizes activities generated by the ordinary light cycle (such as eating and laying eggs) and causes a time mismatch in mutual interactions.

Noise pollution has greatly changed the acoustic landscape and interfered with the acoustic interactions of insects, plus their auditory monitoring of the environment, resulting in huge fitness costs. Ultimately, the impact of electromagnetic contamination on insects plus other life forms (including humans) are still poorly understood and worth further exploration. Other threats comprise of simulated light, which can interfere with their mating rituals. Pesticides that may damage insects or their respective prey; plus water pollution, targeting species with aquatic stages. The report isn't a tally of the globe's firefly populations. However, this is the first time scientists have gathered applicable information, solely based on expert opinions, meaning, what's the key threat that faces fireflies throughout the world. Over the last decade or so, many people have mentioned that they've not been seeing fireflies as much, which could explain the effects of ecosystem contamination (Collins, 2012).

1.8 CONSEQUENCES OF INSECT LOSS FOR ECOSYSTEM FUNCTION

Figure 1.4: Insects vary in form and behavior based on the ecosystem niche that they occupy.

Source: https://bogleech.com/bio-beetles.html

According to the first Global Scientific Review, insects in the world are advancing along the path of extinction, threatening "catastrophic decline of natural ecosystems". The analysis found that well over 40 percent of insect types are declining, and one third are in endangered status. The level of extinction is up to 8 times faster compared to birds, reptiles and mammals. Based on the best existing data, the total number of insects has declined at an alarming rate of 2.5% per year, which indicates that they may disappear within a century or so. The globe is at the beginning of its 6th universal mass extinction event, and large animals are reported to be more susceptible to huge losses. However, insects are generally the most varied as well as abundant animals, outweighing humanity up to 17 times.

Researchers mention that these are the "necessary conditions" for all ecosystems to function properly. They are food for other organisms, pollinators and nutrient recyclers. The collapse of insect populations has recently been described in Puerto Rico and Germany but comments strongly suggest that the calamity is universal. Insect trends approve that the 6th major extinction occurrence is profoundly affecting life forms across the planet. Unless people change how they produce food, the entire insect biodiversity will go extinct within a few decades. At least it can be said that this might have a disastrous effect on the world's wildlife ecosystems.

The analysis, printed in the Bio Protection journal, mentioned that intensive agriculture is among the key drivers of the decline, especially the heavy usage of pesticides. Urbanization ,as well as, climate change are likewise important factors. If the loss of insects cannot be prevented, this may have catastrophic effects on the world's ecosystem and human survival. Through the last 25 to 30 years, the ratio of 2.5 percent yearly loss has been "astonishing" and "rapid". In 10 years, insect numbers will reduce by a quarter; in 50 years they will go down by half; while in 100 years there may be nothing else left.

Among the most profound effects of insect extinction is on various birds, amphibians, fish and reptiles that feed on insects. When this food source has been taken away, then it means these animals shall probably starve to death. Puerto Rico has experienced this chain reaction, besides recent research showed that the number of ground insects fell by 98% in 35 years. Butterflies and moths in particular are the hardest affected. For example, between the years 2000 and 2009, the number of butterfly species widely distributed on cultivated lands in England fell by 58%.

In general, the United Kingdom suffers the most insect decline damage, although this may be the result of more in-depth research than most places. Bees have similarly been severely affected. Only half of the existing bumblebee species discovered in Oklahoma in 1949 existed in 2013. The number of bee colonies in the United States was roughly 6 million in the year 1947, however, it has since dropped by 3.5 million. There are also well over 350,000 beetle types, many of which are believed to have been reduced, particularly, dung beetles.

1.8.1 Insect Success from a Conservation Perspective

Insects have achieved great success as fauna. They control the compositional diversity of all parts except the saltiest and coldest areas of the earth. However, insects are now decreasing at an alarming rate. As far as the lack of the planet is concerned, this is a very worrying and terrible trend. Insects add to the overall maintenance of land and freshwater habitats, the provision of their services and their resilience. The enormous effect of humanity is impacting on this advantage; however, few people understand that life systems upon which they rely are being devastated at a disturbing rate. Action is needed, so are fresh perspectives, if it's necessary to maintain insect biodiversity and eco-services through the modern century.

Insects are among the most special living things on earth, accounting for 70% of all living things. Except for the coldest and saltiest locations, they all dominate. They live in the desert of tropical forests, and swamps as well as huge waterfalls. These are the majority that few people see, and they are hidden in clear views. All land and freshwater plants, including mosses and wormwood, are closely related to insects. Many plants possess flowers, besides their reproduction, relies on insect visitors pollinating them to reproduce. In fact, all lizards and frogs require insects to preserve their population.

In short, insects form the basic context of all land-related ecosystems. In addition, we cannot do without them because they account for one-third of our food, especially the most nutritious ingredients in our food, like fruits and nuts, which depend largely or completely on the pollination of insects. The success of an insect depends mainly on the insect's body system, which are three (head, chest, and abdomen), its extensive skeletal structure, and highly variable physiological mechanisms. The head is equipped with a large number of sensory parts as well.

The chest has high-efficiency feet suitable for different environments, most notably the wings used to disperse and quickly find resources and partners. Consequently, the abdomen hosts a variety of digestion tracts and reproductive systems that spun millions of eggs in certain species. Certainly, the reproductive potential may be rather high. Researchers have calculated that if a pregnant aphid is allowed to breed without mortality, then one year later, the globe will be overrun with 14.7 kilometers deep of aphids.

Insects have a variety of mouthparts that are molded from strong chitin, so they can gnaw, rasp, slurp, and dig through various organic tissues. They can furthermore produce a variety of chemicals used for attacking, defense, and disguise from predators, partner attraction and food absorption. In addition, their sensitivity to some chemicals may be unusual. Some moths can detect companions a few kilometers away, while other moths (such as certain parasites) can detect prey lodged deep in plant matter. Insects aren't only objects; however, they are also interactors.

They are the most closely related ecological species among all living things. A single biome with only 1,000 species can result in 5 million possible interactions. It means that there are more than one million species of insects described, and there may be around five million species in total, indicating that insects interact with every part of the land and freshwater fields.

2

Organizations for Insect Conservation

CONTENTS

2.1 INTRODUCTION

Everything on the globe does have a role to fulfill in its entire lifetime. Living things do have their responsibility, hence the name, living things, for they know it deep within them that they must fulfill a certain role within their lifetime. Non-living things, they too do have a role to fulfill, however, they need the help of the living creatures for their roles to be accomplished, hence the name non-living for they have no willpower within them. All things, be it living things, or non-living things, should be protected, preserved, and well cared for.

To human beings, who are the most evolved living things, not all living creatures are of any significant value to them. However, in our current evolution age, some have been appreciating the existence of previously undervalued creatures. Insects are one among the living organisms which had been greatly neglected by human beings. As stated above, with our new evolutional age, human beings have seen the importance of insects.

With the current appreciation of insects, human beings have ventured into further study of what insects are. Thorough research is being conducted about their habitats, their behaviors, their structures among other things. With the conduction of the research, some insects have been found to be affected by most of our activities as human beings. Large numbers of insects have been found to be facing the danger of ending up being extinct.

The threat of extinction of insects has been quite alarming. This is because insects have been discovered in recent years to be of great benefit and help to us human beings, and to the entire ecosystem. The fear of not having a well-balanced ecosystem has pushed the establishment of various organizations.

The organizations have been aimed to purely concentrate on the preservation of insects. Almost every country has had to establish an insect's conservation organization for the greater benefit of us all. Global organizations have also been set up to help the national organizations in carrying out their different activities of insect conservation.

Before the various organizations were formed all over the globe, there must have been a strong reason that pushed governments and other regions to form them. They must have had a better understanding and a great desire for insects to the extent of being ready to go to such miles of forming bodies that will help in protecting insects in general.

One, insects have been known to be quite good at the pollination of plants. We all, both human beings and other creatures, rely on plants as a source of food. Just as for us human beings, or other animals, we need to combine two cells from different species of the opposite sex so as to produce a young one, so do many plants. However, in plants, the movement of such reproduction pollen grains, which are more of the reproduction cells found in animals from a female plant to a male plant does not happen automatically just like in other living creatures. Plants require the help of carriers to move their pollen grains. Some plants require carriers such as wind, water and insects. However, a large population of plants relies on insects to move their pollen grains from one plant sex to another. The insects happen to attach themselves on the top part of a plant flower as they feed on the sweet nectar fluid found in flowers. The anthers which are the male part of a plant that contains pollen grains are always found close to the nectary so that as the insect sips the nectar fluid, the anthers release pollen grains which will be attached to the body parts of an insect. One flower can never satisfy an insect, so the insects are known to move from one flower to another. Upon arrival to the next flower, the insect attached itself to the stigma, which is the female part of a flower. The pollen grains attached to the insect's body are released into the stigma hence the pollination of a plant which leads to the production of fruits on which human beings and other creatures feed on (Collins, 2012).

To most human beings, insects are known to cause great damages. Some are known to destroy crops, others are known to cause insect diseases to human beings and other creatures. However, we all tend to focus mostly on the negative side than the positive side. That is, we fail to appreciate the fact that there exist carnivorous insects. As the name suggests, carnivorous insects are known to feed in their fellow insects. The existence of the fact that all creatures depend on plants for their survival, be it indirectly or directly does apply here. This is because the carnivorous insects feed on insects which are known to destroy our food crops. Such an insect food chain helps balance the ecosystem by maintaining the numbers of harmful insects at a controllable rate.

We rely on insects for several resources which help us in our day to day activities. Insects such as the bees are known to provide us with several products. The most famous product from bees is honey. Honey is said to be of great value more so when it comes to health benefits. Bees benefit us with wax. Wax can be used in the manufacture of various products as the candles which help in lighting, polish materials more so the shoe polish products, models among other products. Apart from bees, there also exist other insects such as the silkworm. Silkworm is known to benefit human beings by providing them with the silk thread. Silk thread is known all over the globe to be used in the production of one of the best and high- quality silk clothing materials. Some insects are also known to benefit human beings by being of some medicinal use. Somebody parts of insects are being used by most pharmaceutical companies to manufacture medicines which help in the healing of various kinds of illnesses that human beings suffer from (Collins, 2012).

Some communities all over the globe value insects as sources of their identity. Some use certain types of insects to declare the identity of their culture and distinguish them from other cultures.

Insects do serve as a source of food to many. Insectivorous plants rely on insects for their survival. Other animals do also rely on insects for their survival. Some human cultures, do also rely on insects as a source of food.

All living things do die at one point in their lifetime. From the time they were introduced into these plants to the last day, they do accumulate a lot in their bodies, be it tiny or huge bodies. The irony is that; the huge dead things only need the help of tiny organisms to help them go back to almost their original tiny bodies as they were when little. That is where Insecta comes in. Insects help in breaking down the dead living organisms. The end product of the combustion process of living things is said to be of great benefit to the soil. More so, when insects play a part in breaking down plants, the end product is known to improve soil fertility hence quality crop production in our farms.

All creatures always have a place to call home. Some creatures end up sharing the same environment but a totally different place to call home, still in the same location. When it comes to insects, they are of different species which means they all have different habitats. Some Insecta are known to have their home sweet home in the soil. They establish their habitats through boring mechanisms. The tiny holes left above the soil aa the insect crawls deep into its safety, help in the facilitation of clean air into the soil. Basically, insects do help in proper soil aeration hence boosting the health of the soil in general.

Generally, insects are of great importance to the entire globe, and their positive impacts on the globe do out way their negative impacts a great deal.

2.2 THE XERCES SOCIETY

Figure 2.1: The Logo of the Xerces Society.

Source: https://directories.onepercentfortheplanet.org/nonprofit-partners/the-xerces-society-for-invertebrate-conservation

The Xerces Society is one of the organizations which have been formed to help in the conservation of insects. It is a non-profit environmental organization that focuses its energy on the conservation of insects, more so the insect species known to be of great impact to both the biological diversity and the entire ecosystem. Its name Xerces is always in the honor of the extinct species of butterflies, the Xerces blue butterfly species scientifically referred to as the " GlaucopsycheXerces". It was established in the year one thousand, nine hundred and seventy-one by its famous founder, Robert Michael. It was basically a national foundation by the time of its establishment. The Xerces Society is one of the many national insect conservation organizations with it being of the United States region. Its headquarters are located in the Portland Oregon region of the United States of America.

As it is often said that, unity is strength, the Xerces Society does not operate solely. It carries out its activities with the help of other bodies of the United States such as the federal and state agencies that include bodies such as the likes of the US Department of Agriculture among others. For it to be effective in its operations, it is well equipped with experts from different fields such as scientists to help in the study of insects; land managers to locate land areas known to be dominated by different insects species; educators to help in educating insects specialists on the proper ways of preserving the insect ecosystem; citizen of the nation who will help in the general promotion of Insect Conservation among the rest of the citizens who are not yet aware of the importance of insects; applied research experts to help in the thorough securitization of what insects are and their locations; institutions which will help in coming up with ways of educating the public on the importance of insects and why they should be preserved; advocates to support their activities and help them when it comes to legal matters among other experts.

The Xerces Society does have a multiple of different tasks which it must fulfill. One of the roles of it is advocacy. Most individuals do not care at all about the safety of the various insect species. It is the duty of this organization to fight for the safety rights of insects and pursued the government to come up with regulations that ensure insect habitats are preserved. In times of use of insects for medicinal purposes, it is its responsibility to enlighten the government on the best ways it could use that will help in ensuring the insect numbers are still balanced. One way is that the Xerces Society does advise the government to practice the rearing of such medicinal insects in large numbers.

With the rise of the challenges facing insects and the neglecting insects face from the most evolved species, the human beings, who care less about them instead of being the ones to protect them, most insects are at the risk of extinction. It is the duty of the Xerces Society to not only campaign but also help the general public in coming up with ways that will ensure the endangered insect species are captured and placed under maximum care that will promote the existence of them in large numbers hence save them from the danger of being extinct.

Not all people have known of the importance of insects. Some people only consider insects to be a nuisance to their lives. Organizations such as the Xerces Society do exist to help in changing the view most human beings have about insects. Xerces ensures that the public is educated on the many insect species around the globe, more so the insects in their surroundings. After a clear understanding of the insect species, the public is further educated on the importance of each other species. The public is also enlightened on the beat interaction mechanisms that could exist between them and the insects.

As an organization, the Xerces Society does have its ever- ongoing projects. One of the projects that Xerces has been known to give it best is the preservation of the most endangered insect species. This society has gone to the extent of pushing for the rehabilitation of the destroyed habitats of most insect species facing the risk of extinction. The push has been huge for donations that have been made since the projects focusing on manual rehabilitation of insect habitats are known to be quite expensive.

Another of its projects is public education. The Xerces is known to be one of the outstanding organization which has given its best in creating awareness to the general public. Websites have been created and made available to help in educating the public. Most of the websites are known to deal with the importance of insects and how to protect them. Xerces books have been published and sold all over the globe. The Xerces Society does conduct regular seminars in different states of the United States of America and even all over the world so as to have a one on one conversation with the public in the importance of insects and enlightening them on the most effective ways of conserving insects.

Most insects are known to breed in watersheds. As an organization whose main aim is the conservation of insects, the Xerces does know its responsibility is to preserve the breeding grounds of most insects. This is because, with destroyed breeding grounds, there will exist no insects for breeding is the main contribution of the continuation of life among insects. The Xerces Society has various projects focusing on the restoration and the protection of the many watersheds all over our environments to promote the breeding activities of iInsecta hence more insects which means insects are conserved.

Insecticides are one of the main dangers facing most insects. The Xerces Society has various projects which are still underway that major on biological control of insects. This is because, with chemical control, most innocent insects end up dying. Again, chemical insect control does mean most of them will die in large numbers, and just like any other creature, every insect does have its positive impacts and negative impacts, and with their large numbers destruction, the positive impacts will not be experienced anymore. A biological way is the breeding of carnivorous insects in most Xerces Society fields which will be used to control the harmful insects.

One of the main projects the Xerces Society has also been known for apart from the protection of endangered species, it has also been known for the insect pollinators preservation. The Xerces Society has given its best in coming up with a short course that aims at drawing attention to the declining populations of insect pollinators. Such courses have been known to focus on the preservation of pollinators' natural habitats. One of the main activities carried by Xerces in supporting this course is the provision of both physical and virtual classrooms for training sessions. Field training sessions are also conducted from time to time either physically or virtually. Topics being trained are the principles of pollinator biology, the economics of insect pollination, basic bee field identification for bees are known to be the major insect pollinators, land management practices, habitat restorations, plant selection among other topics. The pollinators' short course offered by the Xerces Society has been sponsored by other organizations in the United States such as the Southern Sustainable Agriculture Research and education (SARE), the professional development (PDP) among other organizations.

2.3 THE BUG LIFE ORGANIZATION

Figure 2.2: The Logo of the Buglife Organization.

Source: https://www.linkedin.com/company/buglife

The bug life organization, officially known as the Invertebrate Conservation Trust is a British based nature conservation charity organization. It was founded in 2004 by four key founders, Germaine Greer, President Steve Backshall and the vice president of that period. The headquarters of the Bug life society is located in Peterborough in England in the Nene Park that is close to the Ferry Meadows railway station and the Granary Beefeater Grill. As an organization, it has two famous additional offices in Scotland and the South West of England.

As its official name, the Invertebrate Conservation Trust suggests, its main aim is to prevent the extinction of all invertebrates and ensure that there is the maintenance of a sustainable population of invertebrates in the entire United Kingdom.

The responsibility of the Buglife Organization is known to be categorized into basically three major roles.

One of the roles is that it is responsible for the study of insects. The organization has greatly invested in research activities that help in the provision of information that will promote a clear understanding of different insect species. Insect specialists are quite well educated in the regions of the United Kingdom to ensure that they use the best techniques for the study of insects. The specialists are responsible for providing the most detailed information on how many insect species are in a certain surrounding, the habitats they live in and their surroundings, their structures, their behaviors, their prey, their predators, their breeding grounds and reproduction process and means of production, their day to day challenges among other things that provides a clear understanding of what insects are. After the research, Buglife is also responsible for spreading its information to other experts who do give a hand in the general conservation of insects, be it educators, advocates, institutions among other experts.

The other major role played by the Buglife Organization is habitat management. The organization does major in the study of insect habitats as stated earlier. It does ensure that the experts do have a clear understanding of how Insecta lives in their habitats, how they interact with each for most insect species do tend to co-exist in one environment. Since most habitats are facing the risks of being destroyed due to the increased human activities, the Buglife does focus on coming up with ways to improve them. Apart from habitat management, Buglife also focuses on the enhancement of the insects' biodiversity. Biodiversity means the environment in which insects interact, outsource their food, breed, reproduce among other day to day activities.

After thorough research has been conducted and the insect habitats preserved with the enhancement of biodiversity, the Buglife Organization does have the responsibility of publicizing insects. Insect specialist has the role of publishing books about the different types of insect species, their way of life, their characteristics among other roles. Apart from books, websites are designed by web designers who maintain them at their maximum level of operation such that in times of any errors, the websites are fixed and brought back to operation within the shortest time possible. Billboards are distributed all over the United Kingdom counties to help in captivating the citizens to be more interested in insects. Seminars are conducted from time to time in different parts as the Buglife Organization educates the public on the importance of insects, and ways to conserve them by enlightening them on the dangers facing insects currently.

Just like any other thing in this world has a history, so does the Buglife Organization. Originally, the organization was only known among its founders and the people who had an interest in insects. Not all the citizens of Britain had heard of the Buglife Organization. It is still remained to be a mystery for ages. However, in the early twenty- first century, the Buglife Organization did a tremendous project that promoted its name worldwide. All news headlines reported about the Buglife project. The main project that helped in promoting its name and making it unique among other insect conservation organization was the saving species programmer. It all started as a small project in one county in the United Kingdom. Its success pushed the investors to go beyond one county and get into more saving species projects all over the country. Campaigns were conducted using all means. The online campaigns did help a lot for people all over the world who had the opportunity to be part of the project. Many endangered insect species were saved thanks to the saving species program.

2.4 ROYAL ENTOMOLOGICAL SOCIETY

Figure 2.3: The Logo of the Royal Entomological Society.

Source: https://www.royensoc.co.uk/

Most of the insect conservation organizations have majorly focused on ways of conserving insects from the risk exposing them to extinction.

However, the Royal Entomological Society has been known for its main role of purely studying insects. It has devoted all its energy to study insects so as to provide trustworthy information of what insects are in general and improve the communication between entomologists for they will all be provided with a common reference point for them to base their latest discoveries reducing the conflicts between them. The Royal Entomological Society was founded in the year one thousand eight hundred and thirty-three as the only Entomological society to ever be established in London. Before its formation, there existed several organizations that lead to its establishment such as the Society of Entomologists of London who came together to form the Royal Entomological Society.

The Royal Entomological Society does have its history as do all the other organizations. In the early months of the year one thousand eight hundred and thirty-three, members of various Entomological science departments met and suggested the idea of forming a national organization that would handle issues to do with entomology in the United Kingdom. During that first meeting, a committee was formed that would further discuss the issue and present their final report in the next meeting whose date was set during the first meeting. The report was compiled and presented to the rest of the entomologists of the country by then. It supported the idea of the formation of an Entomological Society in the nation. The officials were elected from the president to the librarian and officers who would help maintain good morning relations with other continental Entomologists. The formed organization, which was then known as the Entomological Society of London started its roles by coming up with a library that had all insect collection information from the period when people developed an interest in the study of insects. The following year, women were allowed to be part of the society, for originally, the Society was composed of men only. After three years of operation, the Society was granted its royal charter by Queen Victoria giving it the privilege of adding the word Royal to its title hence its current name the Royal Entomological Society.

The society does have its authoritative structure to ensure that all its roles are conducted in the most effective way possible. It has a patron, who is and has always been the queen and its vice patron being the Earl of Selborne. Apart from the patrons, it does have a council that is under the rule of the president with other members who have different roles in the council.

The main role of the Royal Entomological Society is known to be the presentation of information on Entomological science in the easiest way such that anyone willing to venture in the study of insects is able to understand their books with ease. The information is presented through publications which have been sold all over the world, regular scientific meetings which are even being conducted all over the globe, supporting and funding Entomological expeditions which are held almost annually in different parts of the United Kingdom, and regular public events some of which are virtual.

Currently, the organization has been able to establish and perfectly maintain its major Entomological library which is located at the society's headquarters in St Albans, United Kingdom.

As a society, the Royal Entomological Society is made of more than fifteen special interest groups that cover different scientific fields within entomology and they all come together frequently to discuss different matters concerning the society. Apart from the special groups, the Society has managed to be in partnership with more than sixty other organizations. The partnership has provided it with enough support to carry out biennial meetings with the public across the United Kingdom. This meeting enlightens the public on the importance of insects and entomology in general. Apart from the biennial meetings, there are also biennial insect festivals which are conducted to celebrate the existence of insects and the Entomological science in general.

Since the Royal Entomological Society is almost a pure study organization, it has majored in the provision of platforms that will present information to the general public. The Society publishes periodical journals which include; The Agricultural and Forest Entomology; Ecological Entomology; Insect Conservation and Diversity; Insect Molecular Biology; Medical and Veterinary Entomology; Physiological Entomology and the Systematic Entomology. Members of the society do receive these journals every time they are published. Apart from the journals, the Society does alas publish a series of different handbooks all of which deal with the identification of insects for it is important for insects to understand what insects are really before venturing into ways of conserving them. The identification of insects enlightens the public on the morphology of insects, biological structures of insects and their distribution in the United Kingdom.

With the help of the Internet, the Royal Entomological Society has been able to grow to international levels. Through its website, the society gives individuals from all over the world to be part of it. It has an international membership platform on the Internet which is open to individuals who are willing to contribute willingly to entomology. The applications are forwarded to a committee of a council that handles the approval of the applications. Worthwhile applicants are allowed to be part of the society after thorough scrutiny has been conducted in them either physically or virtually depending on the application means they used. After approval, the new members are given the privilege to make use of the title Fellow of the Royal Entomological Society more so in events which they attend that have anything to do with Entomology for Royal Entomological Society is a well-known organization all over the globe.

The Royal Entomological Society is a well-known organization all over the globe. Its name can be recognized basically anywhere even in situations that have totally nothing to do with entomology. Because of its popularity, the society has earned itself different awards. It has also opened a platform for project contributions from individuals all over the world. Winners of its projects are awarded depending on the category they have participated in. Some of the awards awarded by the Royal Entomological Society include; Royal Entomological Society Goodman Award; March Ward for Insect Conservation; Alfred Russel Wallace Award; J. O Westwood Medal and the Wigglesworth Memorial Lecture.

Just like any other organization, the Royal Entomological Society does have a badge. The badge is its unique identity which is used to identify it and represent it even without the need of using its name. Its badge does have the phrase "Royal Entomological Society "in it spelling out to the world that no one can ever use the badge for other purposes apart from the representation of the Royal Entomological Society.

2.5 BUMBLEBEE CONSERVATION TRUST

Figure 2.4: The Logo of the Bumblebee Conservation Trust.

Source: https://www.justgiving.com/bumblebeect

The Bumblebee Conservation Trust is a unique organization known to major its efforts mainly on one insect species, the Bumblebee. It is basically an organization in the United Kingdom that is known to focus its effort on the monitoring and conservation of Bumblebees and the management and enhancement of their habitats. It was formed in the year two thousand and six under the Registration of a charity organization in the United Kingdom. Its headquarters are located in the Beta Center in the University of Stirling Innovation Park in the United Kingdom with its original chief executive officer being Sir Gill Perkins.

The trust does have its own history. It was first established by Dave Goulson in the year two thousand and six who obtained its foundation money from the Heritage Lottery Fund back them. The organization decided to focus purely on Bumblebees because their population in the United Kingdom was dropping at high rates exposing them to the risk of extinction.

Bumblebee Conservation Trust has three major roles to play. One of its major roles is to ensure that Bumblebees in the United Kingdom are well taken care of. Not all parts of the United Kingdom are facing the danger of having extinct Bumblebees, however, the trust does insist that all of the Bumblebees all over the UK, whether rare or common, be conserved.

Its second major role is to educate the public on Bumblebees. Just like most of us do not value insects, the Bumblebees are not valued for they are part of the Insecta class. The trust ensures that the public is made aware of the existence of Bumblebees. They are educated about their appearance, their structures, their behaviors, their distributions and their benefits. Bumblebees are known to be of both environmental and economical benefits hence the public should be enlightened on their importance to them.

Most organizations do face many challenges that might lead to their collapse. The Bumblebee Conservation Trust has made it their priority that they should always be ready to face any challenge that comes their way and they tackle it in the most effective way ensuring that it does not affect their general performance.

As an organization, the trust has managed to come up with different projects all of which have been aimed at conserving Bumblebees. One of the projects that promoted its name all over the globe was the project on national Bees for Everyone which was majorly focused on educating the public mostly on the threats facing Bumblebees and enlightening them on ways in which they can actively help in conserving their existence by safeguarding, restoring and even creating new habitats for Bumblebees. This project was supported by people all over the world and encouraged the establishment of even more projects all over the globe all aiming at conserving Bumblebees.

As stated earlier, the Bumblebee Conservation Trust is a charitable organization that has been registered with the office of the Scottish Charity Regulator and Charity Commission for England and Wales. It has several members most of whom are from the United Kingdom with only a few of its members being from other countries. However, with its growing fame levels, more and more foreign members are joining the Bumblebee Conservation Trust. Apart from its main headquarters which are located at the University of Stirling, the trust does have several offices all over the United Kingdom.

2.6 BUTTERFLY CONSERVATION

Figure 2.5: The Logo of the Butterfly Conservation society.

Source: https://twitter.com/savebutterflies

Butterfly Conservation is a United Kingdom wildlife charity organization that was formed to majorly concentrate on the conservation of butterflies, moths and the environment in general for the in the environment we have the habitats of these insects. Generally, the butterfly conservation group not only focuses on butterflies but also on the entire insect species of the Lepidoptera order of class Insecta. Most of Lepidoptera insects are known to reside on land hence they tend to be exposed to many human activities that affect their existence. The organization has played a major role in participating in many projects all of which aim at the conservation of butterflies and moths all over the United Kingdom.

The Butterfly Conservation society is known for its motto "saving butterflies, moths and our environment ". It was formed in the year one thousand nine hundred and sixty-eight as a non-profit organization whose main purpose was to improve the environment for mainly butterflies and moths. Its headquarters are in Manor Yard in the East Lulworth Wareham Dorset in the United Kingdom. It is said to have over fifty thousand members most of which are from the United Kingdom. The most recent president of the organization is said to be Sir David Attenborough with its Chief executive officer being Julie Williams. As a big organization, the Butterfly Conservation society is composed if offer branches all of which are well spread all over the UK.

The organization is said to have been formed in the year one thousand nine hundred and sixty-eight after few naturalists met together and discussed the importance of butterflies since they beautify the environment and moths since butterflies are known to have evolved from moths in the life cycle of insects. As stated earlier, the organization was registered aa a charity organization later in the same year. It is known to operate the three main and world's largest butterfly and moth recording schemes for it is said to have over sixty million records on butterflies. As an organization, it does have an annual journal that updates both its members and the general public on its progress in conserving butterflies, moths and the environment in general.

One of the major roles of the Butterfly Conservation Society is to enlighten the public on the importance of conserving mainly butterflies and moths. The society emphasizes the fact that butterflies beautify our environment and they do help in easing the tension among many individuals more so in gardens where people go to relax and get some fresh air. The public is educated on the fact that butterflies are quite sensitive in nature and they do help us in assessing the changes in our environment for a slight change that does reflect on butterflies and moths. The public is also made aware of the fact that, in the previous years, a large population of both butterflies and moths have been observed to decline at high rates.

Its second is the maintenance and enhancement of different landscapes for both butterflies and moths are known to dominate in well- managed landscapes. Butterfly Conservation society is known to provide almost all landowners with the well -detailed information on ways in which they could manage their lands to ensure that the habitats for butterflies and moths are well preserved and do not end up being destroyed.

Regular research activities are conducted from time to time to help in updating the previous records on butterflies and moths. Information obtained from the research is also important for it can be used as a guide in the conservation of butterflies and moths. The organization does have reserves almost all over the globe but more so in the United Kingdom all of which help in conserving the environment, butterflies and moths.

Apart from educating the public on the importance of butterflies and moths, the Butterfly Conservation society does also provides a platform for individuals who are quite interested in the study of butterflies to join a training session and be educated basically on all, there is to know about butterflies.

Just like any other organization, the Butterfly Conservation society is composed of a president, vice president and other committee members all of which play a major role in ensuring that the organization runs effectively.

2.7 AMATEUR ENTOMOLOGISTS

Amateur Entomologists society is an organization based in the United Kingdom and it was formed for all people who are interested in insects, be it the study of insects, or basically anything that has to do with insects. It was formed in the year one thousand nine hundred and thirty-five. Originally, it was referred to as the Entomological Exchange of Correspondence Club. One of its major roles back then was to provide a platform for all entomologists and insects amateurs so that they could be able to not only share their findings but also have a change of acquiring any reference materials they required.

Two years after the society was formed, it changed its name to Amateur Entomologists for most of its member where individuals who were well informed about insects and were willing to go to any mile in search of a better understanding of insects.

The society has devoted itself to production and publication of annual journals all of which provide information on insects. However, most of the journals preach the gospel of insect conservation and the importance of insect conservation. One of its most known journals all over the globe is the Entomologists Record and Journal of Variation that covers all the information to do with insects from their structure to their conservation ways.

Most of the publications are sold through different online marketing platforms. The reviews to such publications are freely made available on the Amateur Entomologists Society. Apart from websites, regular seminars are conducted from time to time all over the United Kingdom with most of them being done virtually due to technological advancement.

The communication about any events that are the Amateur Entomologists society is planning is conveyed via the several social media platforms which are run by different individuals who are specialists in those areas. Amateur Entomologists society has an account almost in every available social media platform creating a one on one interaction with their members and anyone interested in being part of them.

2.8 THE BRITISH DRAGONFLY SOCIETY

Figure 2.6: The Logo of the British Dragonfly Society.

Source: https://www.charityjob.co.uk/recruiter/british-dragonfly-society/43285

The British Dragonfly Society is an insect conservation Society that majors its operations in the United Kingdom. It focuses on the conservation of Dragonflies and Damselflies. It was founded in the year 1983 with its founders being Sir Rod Din and Bob Merritt.

As the name suggests, the society has devoted itself to the study of dragonflies and its habitats. The society does also campaign for the preservation of these two insect species all over the United Kingdom. It does also educate the general public on the importance of Dragonflies and Damselflies and enlightens them on ways in which they could promote the conservation of these insects.

As an organization, the British Dragonfly Society runs the largest recording group about Dragonflies in the world which is referred to as the Dragonfly Recording Network which collects information on dragonflies and other insects belonging to the Odonata order of class Insecta.

Apart from educating the public, several projects are still underway to help in the protection of basically all habitats of the insect species belonging to the Odonata order but more so the dragonflies and the Damselflies.

2.9 LOST LADYBUG PROJECT

The United States of America entomologists thought of a way of engaging young children in insect conservation by coming up and forming the Lost Ladybug Project organization. Lost Ladybug Project is a non-profit organization that has devoted itself to educating young children about scientific information with the entomological science being one of the sciences passed on to children.

The organization equips the children with unique confidence when it comes to matters to do with scientific arguments. The children are also able to compete in different scientific projects. Well informed children are able to identify themselves personally with science without the fear of being mistreated. This organization is mainly helping the society in developing a generation of adults who can identify themselves with science and engage in scientific discussions, promote scientific policy -making and engage in scientific thinking.

The organization was formed in Ithaca states in New York as a non-profit organization that focused on purely conservation projects and education. Its headquarters are located in Ithaca New York state in The United States of America.

The Lost Ladybug Project was formed in the year two thousand by scientists from Cornell University who had previously embarked on a mission to study the ladybug insects. During their study, the scientists discovered that ladybugs where facing some type of challenges that risked their extinction. With this information, the scientists made it their mission to study more about the ladybug insects and educate the public about them.

One of its unique characteristics is the fact that these scientists chose to not only educate the adult about ladybugs, but they chose to make it be more of a scientific project for young children allowing them to know more about insects and have a good upbringing being well informed about insects.

This would ensure that, in years to come, the adult generation would be quite responsible for insect conservation and environmental conservation in general. That is why the organization, not only focused on enlightening young ones about the ladybug insects, but they made it their mission to ensure that all scientific information was passed on to children.

The Lost Ladybug Project apart from focusing on educating the public and majorly the children, did also focus on the study of all Ladybug species to ensure that they are well conserved and their habitats well- taken care off.

2.10 POLISH ENTOMOLOGICAL SOCIETY

The Polish Entomological Society was an organization that was founded in the year one thousand nine hundred and twenty-three. Different Entomologists of those years came together and discussed the importance of having a common organization that would serve as the central reference point for most of their activities. Just like any other motion in any debate meeting, the idea was not supported by all but a large number of the Entomologists did suggest a committee be formed and scrutinize the idea. After a thorough discussion was done by the committee, a final meeting was held to present their report. The report seemed to support the idea of having a common society hence a large number of the entomologists present in that meeting said yes to the idea and the Polish Entomological Society was formed.

The main roles of the organization were to support the different activities which are frequently conducted by entomologists. It was to help in the conduction of thorough research in insects in all branches be it biological research on insects, taxonomic groups of insects, applied entomology among other issues to do with the insect study.

Since the Polish Entomological Society is more involved in study activities, it does leads to the publication of different journals that will educate the public on insects. Some of the published journals include the likes of the Polish Journal of Entomology journal, Entomological News journal, Keys for the identification of Polish Insects journal, Polish Entomological Monographs journal among other journals.

Most of the journals are updated periodically and made available to the public. Access to these journals is either through the online platforms belonging to the Polish Entomological Society.

2.11 ENTOMOLOGISCHER VEREIN KREFELD

Figure 2.7: The Logo of the EntomologischerVerein Krefeld.

Source: https://en.wikipedia.org/wiki/Entomologischer_Verein_Krefeld

The EntomologischerVerein Krefeld is an insect conservation society that is located in Krefeld, North Rhine Westphalia region in Germany. It was founded in the year one thousand nine and five by the famous entomologists back then in Germany, Sir Martin Song. The main agenda in his mind which lead to the formation of this society was to ensure that it would serve as a collection center for all information that has anything to do with insects and also helps in the preservation of insect specimens collected for study purposes. Basically, its main role is that of maintaining records of insect information that has been collected from all parts of Germany by several entomologists to ensure that they all have a central reference point.

The Society remained to only be known in Germany until the year of two thousand and seventeen when it conducted a special study that made it famous all over the world more so among insect amateurs. The society published a report on its recent study that stated that the population of flying insects was greatly decreasing, not only in Germany but all over the globe urging the public to be more keen while conducting their day to day activities so that they save the family of flying insects from extinction.

The membership for this society is mainly composed of individuals who are more of insect amateurs for not all of its members are entomologists. Some of its members happen to have made society more of a place where they can practice their hobbies of preserving their most valued insect species. The membership number is not that high. This is because the EntomologischerVerein Krefeld society is more of a sub-branch of the German Association for General and Applied Entomology. Despite the fact that society is more of a sub-branch of a major organization, it does open its doors to new members. Interested individuals submit their application either virtually or physically. The applications are later scrutinized and the worthwhile applicants approved to be members of this society to help in the study of insects.

Apart from the two thousand and seventeen publication that promoted the EntomologischerVerein Krefeld society, the society has also participated in the publication of other reports and journals all of which have been said to be quite interesting to the general public more so to individuals who have always had some kind of interest on insects. Most of its publications are advertised via the website with a review of them being provided on the Internet hence encouraging members to purchase them. Apart from the website, the society does have accounts in almost all of the social media platforms. These accounts are run by specialists who help in posting updated information about the recent activities the society is about to conduct.

3

Increasing Public Engagement in Insect Conservation

CONTENTS

3.1 INTRODUCTION

There are a variety of reasons why people need to conserve the various species in the environment, this also includes insects. The conservation reasons can be grouped into categories for instance our survival and economy depend on the various species that also live on the planet. They are a source of tourist attraction and increase the aesthetic value of the environment. We also a moral obligation to act and show compassion towards other species and life forms that exist in the world.

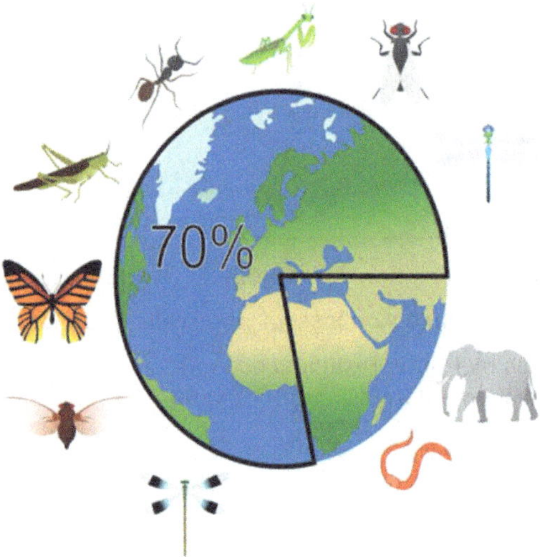

Figure 3.1: Insects make up the largest part of animal biodiversity in different terrestrial ecosystems and aquatic ecosystems.

Source: https://www.intechopen.com/books/insect-science-diversity-conservation-and-nutrition/insect-conservation-for-the-twenty-first-century

Is the conservation of insects necessary or a concern? They are various reasons why we need to conserve insects. For starters, they make up about four- fifths of all animal biodiversity on the earth. As much as people have tried to conserve species with the threat of extinction, over the years, insects have had little attention pay to them by conservationists. We cannot say we conserving biodiversity if we exploit and ignore species that make up a large percentage of the animal population (Cheesman& Key, 2007).

Apart from the fact that insects having a large population and possessing a variety of species, they are important in the survival of other species such as plants, birds and mammals. They also play an important role in enabling human survival by providing ecological services they act as pollinators, a source of food, they are used in research, they are a source of raw materials, they are Important in the field of medicine and they can also be used as biological control of some pests. These are just but a few reasons.

Some of the services provided are indirect but are none the less important. For example, insects are important in the recycling of organic matter of plants and animals. Life as we know it would not exist without such processes for recycling. Insects in general have been able to conquer different habitats and also adapt to such environments. As a group, they have the probability of surviving on this planet and similarly support other species but many individual species are vulnerable to rapid changes as a result of human activities and influence. This makes their survival impossible in different localities which others would be suitable habitats for them. The problem is even greater for species with a small mobility range and therefore are unable to colonize or recolonize environments following the chance of extinction due to sudden or short term events such as weather conditions that are extreme such as long periods of drought, floods, hailstones and many more. Due to such events, species of insects become vulnerable to extinction across large areas, this may be at a national level or transnational. Due to the small size of insects, the extinction or vulnerability may go unnoticed.

Insects on their own can take care of themselves and adapt to new environments and this has been observed in areas that have not been encroached or inhabited by humans. Other than human beings degrading insect habitats, insects face other threats such as pollution, over application of pesticides and wide- spread urbanization which has led to collapsing of numerous insect populations. This means that human beings destroy and isolate suitable habitats of such species depriving them of a means and chance of survival. Conservation of insect species means that we need to identify, stop and reverse the harm we do in destroying their environments (Cheesman& Key, 2007).

There are different ways of conserving insects and their habitats. The key to insect conservation is the protection of their habitats and enhancing them as found necessary. Habitats can also be created throughout a landscape such that even the least mobile species can find a suitable habitat for their survival even in unstable conditions. Although insect habitats can be conserved, other substances such as pesticides can harm insects and other forms of pollution. Pesticides are used in the agricultural control of pests but sometimes harm species that are not targeted of which some of them are beneficial to farmers.

Using integration can be a method of pest control, this means using different methods to ensure they do not reach in non-target habitats. Research can also be done in order to protect species that are vulnerable to extinction in order to avoid unnecessary restrictions. Unlike vertebrates which are found in small numbers, insects are found in large numbers and some of the insects are less likely to become extinct.

Several organizations are working to ensure certain habitats and species of insects are conserved. Entomologists who study insects are some of the personnel involved. They have knowledge of insects and their habitats which is important for anyone offering advice on land management, including nature reserves and other designated sites. They can also provide information to amateur entomologists and the general public on the importance of conserving insect habitats throughout landscapes. One of the best ways of doing this is by becoming members of local wildlife trust. Such organizations welcome people who can lead entomological field meetings or give talks on insects. Effective insect conservation cannot be done without doing the following having knowledge on biodiversity, understanding the past and present distribution of biodiversity, implementation of monitoring systems that are suitable for arthropod populations to disentangle stochastic and natural variation as a result of human activities; identifications of factors that are harmful influencing arthropod populations and cascading effects on services provided by the ecosystem and finally seeking strategies to alleviate the action of harmful factors and restoring ecosystem services (Colla, 2017).

Despite the importance of insects, it is only recently that they have been included in biodiversity conservation. This is done at different levels of operation from species- level through to the level of landscapes with a majority of decisions being made by a scale of nation -states through national biodiversity and strategy action plans. There are also initiatives at a global level.

In the 21st century, insect conservation is seen against six themes that are inter-related. They include philosophy, research, policy, psychology, practice and validation. These are done against a background of species, landscape, national and global level of conservation operations. This is done to hasten the initiatives that have been put in place to save the current diversity in the insect population.

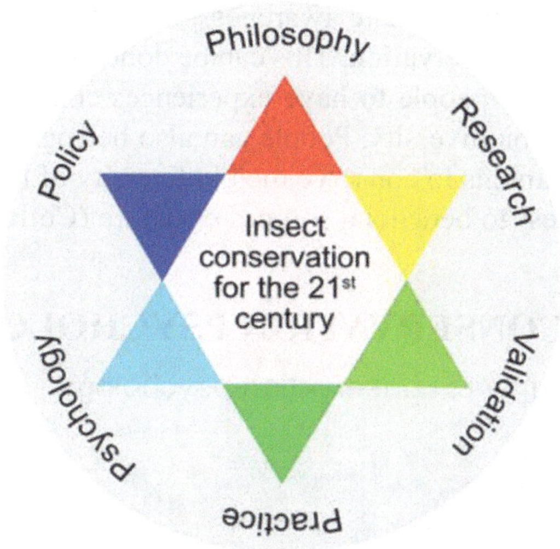

Figure 3.2: The themes that surround the conservation of insects in the 21st century.

Source: https://www.intechopen.com/books/insect-science-diversity-conservation-and-nutrition/insect-conservation-for-the-twenty-first-century

People seek their connection to nature and this is evidenced by visits to national parks in large numbers annually. One of the biggest challenges in conservation is the lack of appreciation of the benefits we get from nature by caring for them this is inclusive of insects. The relationships between humans and nature for example insects is complex which the brain is not equipped to deal with. Sometimes even having a scientific understanding of the importance insects have on the survival of other species, we might not still value them. However, this is changing for example our understanding that simple and small insects such as bees are responsible for processes such as pollination which then goes to affect food security locally and globally. In order to improve conservation awareness and action, it is necessary to first engage insect conservation psychology. Citizen science and red listing play a top role in doing this. The use of insect conservation psychology can be used to impart a culture of personal and social responsibility and this creates political will which drives the conservation from being theoretical to being actualized at the local, national and global level. Understanding the psychology of people in relation to insect conservation will help stakeholders involved such as conservation agencies, governments on how to better engage the public in order to achieve their objectives. Organizations can use education and publicity to create awareness. People can also be engaged in activities to do with conservation. This can be done by developing programs that are educative for people to have experiences connected to nature and also get lessons on biodiversity. People can also be engaged in establishing and maintaining habitats to conserve the biodiversity of insects and also the facilitation of access to beneficial sources of nature (Colla, 2017).

3.2 INSECT CONSERVATION PSYCHOLOGY

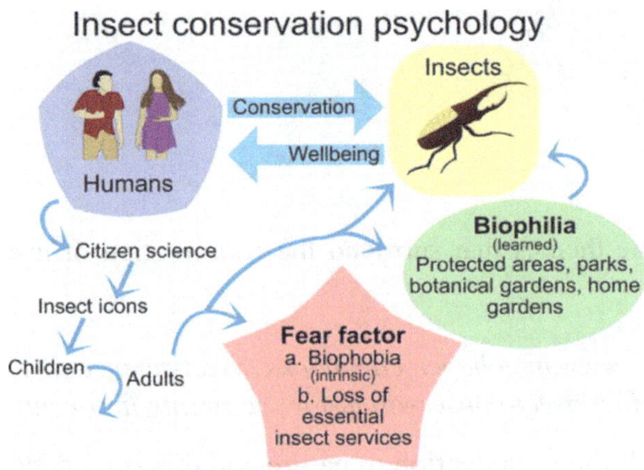

Figure 3.3: The relationship between insect conservation psychology and insect conservation.

Source: https://www.intechopen.com/books/insect-science-diversity-conservation-and-nutrition/insect-conservation-for-the-twenty-first-century

3.2.1 The Need for New Psychology

Psychology is the study of human behavior while conservation psychology recognizes the dependence of human beings on nature not only the provisions but understanding why and how they are essential for our wellbeing. It expresses the process of caring for nature and nature in similarly providing its resources to us. It helps us to understand the connection people have to nature and how to better engage them in conservation activities (Didham et al, 2020). People are not doing a good job especially in the conservation of insects. The view of people on damage to biodiversity will contribute significantly to the efforts that will be put in place to prevent the death of species. It can be perceived in three ways like the loss of beneficial assets, massacre by interspecies and evidence that humankind is toxic to the biosphere. Humans should start by acknowledging that they are an origin of destruction that is presently there and is also observed towards other species including insects. The damage is attributed to social change and advancements in technology. Conservation of nature commands the presence of psychological instruments which is vital in changing our mentality and find new and innovative ways of conserving insects.

With the progress in research, people have a developed a better mastery on weather and a grasp on the management of diseases unlike the past when it was difficult but there is an absence in the comprehension of the importance of insects in both terrestrial and fresh water ecosystems. This can be attributed to our lack of appreciation given their small size.

Most environmental policies are created around the instrumental valuation rather than intrinsic valuation and therefore has more currency in considerations made in relation to conservation. It is used in ecosystem service valuation and arguments from conservationists are that nature should be conserved also on the basis of its intrinsic value with a deep understanding and reflection on what is good, worth protecting and how we are supposed to interact with the world that surrounds us (Cheesman& Key, 2007).

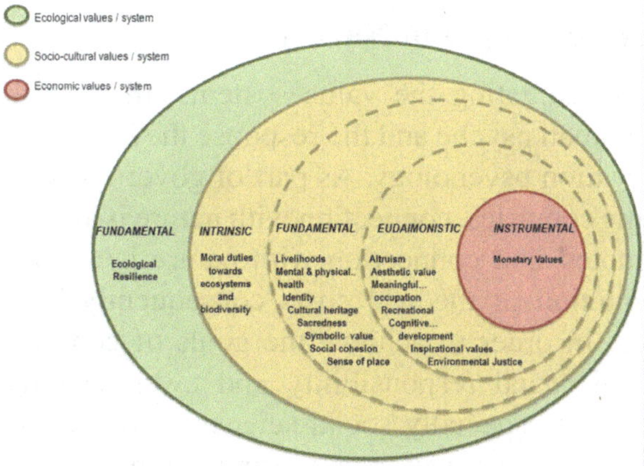

Figure 3.4: Different domains of valuation used in the classification of ecosystem services.

Source:https://www.researchgate.net/publication/312303940_Widening_the_evaluative_space_for_ecosystem_services_A_taxonomy_of_plural_values_and_valuation_methods

Our value of nature also needs to be consolidated into an action plan that is more cohesive and this is found within conservation psychology, which is also found within environmental psychology. This broader area of study that is environmental psychology, aims to provide tools that are used in the facilitation of environmental management and governance. This goes beyond protecting species and the engagement of public audiences and stakeholders so that there is the delivery of objectives used in the management process which are used in the achievement of societal and ecological goals.

There are several dimensions to the valuation approach that leads to action towards conservation. It reconciles the perceived dichotomy between instrumental and intrinsic valuation. It brings forth a range of related values in the case of the interrelationship between human beings and nature. With regards to insects and their complexity together with the knowledge gap on them, there is the need to also integrate both scientific and an ethical approach in their conservation. Conservation action has biological levels. It ranges from the conservation of evolutionary significant units, through the maintenance of networks of insects that carry out pollination, to the conservation migrations that occur in large numbers. It also includes problems at a global scale such as the fragmentation of habitats which is one of the largest threats to invertebrates (Hobbs, 2012).

3.2.2 Connecting People to Nature

In addition to investigating the value systems, there is the need to also understand the human psyche and the response they have towards nature for example conservation psychology. As part of government health sponsored programs in some countries, connection with nature is already been executed as it has been noted that connection with nature is a strategy that can be used in the reduction of medical costs consequently being beneficial to local economies. In order to achieve the goals of conservation, scientists have incentivized social responsibility and awareness using tools from other disciplines but especially social sciences. The connection of people with insects using nature experiences, caring for insects by individuals and society at large can be achieved. This process can also be referred to as rescuing the extinction of experience and is significant to different groups for example who do not relate well to a large number of insects but are also the pool from which future insect conservationists will emerge from.

Figure 3.5: A person snorkeling comes encounters a jellyfish.

Source: https://blog.nwf.org/2013/07/national-wildlife-photo-contest-category-spotlight-connecting-people-with-nature/

3.2.3 Empathy, Caring and Insect Conservation

Satisfaction from humans is linked to greater feelings of empathy and care. Connection with nature promote personal aims and choices that reflect caring and feelings of good health. Positive feeling includes the care of nature which supports beneficial decision making for the environment. It can also promote caring for other people especially communities that are disadvantaged. With changes in technology, people continue to interact less with nature and similarly understand less. It suggests that the psychology behind moral behavior is not complex. This means it has the possibility of having an influence on the success of conservation activities. Public opinion and also demand are tools that have power on the conservation cause and are more likely to engage decision- makers and compel them to act (Hobbs, 2012).

3.2.4 Should we like insects?

One of the greatest challenges of improving insect care is that many insects are not nice, or useful and are ugly. It bears the instrumental vs. the intrinsic valuation debate. The human-nature relationship is complex but it can be intuitively typified, using an integrated framework of values, which are also known as biophilia values. An important fact is that they are learned and therefore are not intrinsic, in contrast, biophobia is intrinsic.

Biophilia is associated with a positive feedback mechanism. Looking at images of nature can be of benefit to the health of human beings and this makes the process of learning a bit easier. The imagery of insects is used on clothing, utensils, fabric and this makes the impression that we do like some of the insects and we spend money on such objects with insect images that adorn ourselves and our homes. This is also passed down to our children. This can be used as an indirect means of creating awareness on the value of insects which can lead to the progression of our conservation efforts and can be termed as a small step in the right direction (Hobbs, 2012).

Kellerts framework on biophilia values is integrative, that is, a spectrum of values from utilitarian to negativistic. Intrinsic valuation is not at the end of the spectrum with physical control and fear, aversion being the endpoints. His typology does not view utilitarianism as an intrinsic value that is opposite but includes both of them as part of the human- nature relationship. Norton rejects the dichotomy of opposites humans' values from values that are self-directed and consumptive. This includes the aesthetic value, spiritual and valuations which are non-instrumental. This means the framework of biophilia lends itself to a universal ethic for conserving biodiversity, with insects as the species that form the largest part of the animal population. An important part of ethic is that different people will have different attitudes towards different situations. The framework will therefore not be about categories but also categories that experience changes that are dynamic (Hobbs, 2012).

3.2.5 Thinking Globally, Acting Locally

A combination of the conservation of nature and human wellbeing is a driver needed for taking conservation to action. It also gains traction towards meeting sustainable development goals. Accordingly, new metrics of human and ecosystem well-being are needed that combine insight from psychology to ecology to describe how the complex of natural, social, human and built assets can contribute to sustainable well-being, with the best system aiming at achieving the goal of a prosperous, equitable and sustainable world.

3.2.6 Our Perceptions on a World that is Changing

Instigation of effective political agendas is also dependent on human will. Most of the time people do not consider environmental protection and resource as not being important. According to Kellstedt, giving reports which are statistical on the decline in nature is not an effective way of stimulating people into conservation action. It happens regardless of how people value nature. The explanation to this is by understanding the human psyche. Understanding our psyche better, conservationists could send out messages that are finely tuned about the need and value for the conservation of insects in order to have positive outcomes. Negative messages for example those sent via different media platforms can elicit defensive responses and lower the openness of people to new information. They are behavioral responses that reduces tension. In other circumstances, this can be coupled with other defensive reactions such as reactance and denial and so have an even more negative effect. The pollinator crisis has stimulated huge efforts in correcting it in many parts of the world. This has also increased concerns in relation to pesticide use. In addition to this, there is also the realization of our health and food source being compromised. The pollinator problem focuses on two major points within our psyche for example the bee. For example, it easy to associate bees with their value and their potential to be used as ambassadors for conservation to people when seen buzzing across a landscape is great. Another spectacle is the migration of monarch butterflies and others (New, 2012).

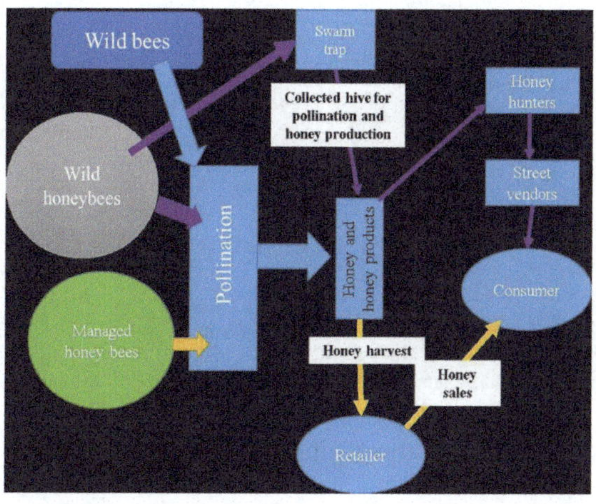

Figure 3.6: The value of bees.

Source: southafrica.co.za/af/die-belangrikheid-van-heuningbye.html

3.2.7 Value of Protected and Green Areas for Human Well-being

Worldwide, eight billion people visit protected areas annually. According to McCarthy, the cost put into conservation of biodiversity is small in comparison to the potential it has in related goods and services provided by biodiversity. Nature provides enjoyment and human well-being as a consequence. This opens a way for awareness to organisms in nature which people do not even have the time to notice. It fits well with the idea of services provided by the ecosystem should be seen as ecological processes and their effects that humans appreciate.

Urban parks, botanical gardens, and dedicated insect reserves play a role that is important for the well-being of humans and are arenas that are important for the purposes of formal learning. Insects that are charismatic such as dragonflies are referred to as public dancers for an appeal to the public. The benefits they provide, suggest that there is underinvestment in conservation. The large number of people, who visit protected areas implies that there is a lack of a dichotomy in values as people are willing to spend on nature and they feel good by visiting such areas (New, 2012).

Figure 3.7: An urban back in the United States.

Source: https://www.nps.gov/subjects/urban/why-urban.htm

3.2.8 Challenges beyond our Control Overpower Our Individual Values

Surveys done in some countries show that people have a relatively high awareness of environmental issues globally but their behavior did not match this awareness. This means while individuals may hold high values their actions are not friendly to the environment, regardless of the environment. Given the decline in insects and their ecosystem services, it is also important that the need for insect conservation effectively elicits both support that is either financial or non-financial from the general public. For example, the Xerces society in the United States; bug life and butterfly conservation in Europe.

These organizations have had outcomes that are positive outcomes from their activities and this contrasts with the way many respond to environmental issues like climate change and there is no likelihood that they will feel responsible for global warming. There are complex and multifaceted reasons for this, but partly insects are more tangible and immediate and they are items we have all experienced in our life both the positives and negatives. Humans lack the moral judgment to identify changes in climate as an important moral imperative. The issue of climate change most of the time does not motivate people to make urgent changes as it is a complex issue and it is also not tangible.

Markowitz and sheriff gave six reasons why climate change does not register in our minds and at the same time, it provides possible solutions to the problem. For starters pushing action against climate change may not work like businesses. Use of economic incentives to motivate people can lead to the creation of conflicts that have a negative relation to materialistic values and environmentalism. Promotion of materialistic values prevents people from developing inherent values that lack a material basis that outdo materialistic values.

When material values are removed from the picture, People lack the motivation to continue with the behavior for its own sake also goes. The most powerful motivation that influences behavior is people's perception of right or wrong. This means that social personal moral norms play an important role in driving pro-social and pro-environmental behavior (New, 2012).

People sometimes view climate change as a truth that is inconvenient but at the same the time they don't view it as being in their backyard. In contrast, the pollinator crisis is also a large scale but it is also tangible. For example, we see people killing bees yet they are creatures beneficial to us, so our instinct will be to protect them. It is also up to conservationists and policymakers to tap into these insights in order to deal with the problems faced while conserving insects. Presently the services provided by insects are being promoted together with nature conservation for example agroecology, which focuses both on pollinators as well as other service providers such biological agents of control. Yet insect conservation especially the services they provide is not equal to the conservation of rare species and those that are threatened mostly, especially with human mitigation for wildlife- friendly service provision sometimes being at odds with the conservation of insects for its own sake.

Although there are major threats to biodiversity, such as habitat loss, invasion by novel species and pollution are challenges that can be observed and this has become global problems similar to climate change. The impacts are also often adversely synergistic which changes to the climate which are mostly caused by different human activities. Biodiversity threats are complex, large scale and not intentional and provide the same kind of individual behavior and moral judgment similar to climate change with reduced concern for conserving the environment. There are increased expectations, with an increase in single species awareness in relation to conservation and recognizing that the protection of specific types of habitats is both tangible and essential (Cheesman& Key, 2007).

Overtime there are areas where awareness in relation to insect conservation has increased. They are the red list, citizen science and the International Union for the Conservation of Nature. Once an insect is placed on the red list, it becomes iconic. A few species of insects that previously were seen by a few individuals become high profile and often become the subject of conversation and action once it is placed on the red list. Insects that are red- listed also allow the protection of certain landscapes as their presence, especially if the species are rare and are found in the same location if it is also not replaceable.

Citizen science is the involvement of informed and enthusiastic sectors of the public sector for recording the distribution of species and engaging them in conservation activities. Citizen science is a field that can be incorporated into strategies dealing with insect conservation. There are different requirements for practices that include citizen science to be included. Some of these include the availability of good guidelines for identification of species we can focus on and the common names for insect species, both of which are necessary, together with support from professionals to identify the species of interest. Capturing images, especially making them readily available to those using digital phones also plays an important role in the creation of awareness of insect conservation. Citizen science has led to several improvements such as helping people to understand butterflies in the United States, stag beetles in Europe and pollinators in the United States and more. The point of engaging and involving more people widens the opportunity for people to acquire more knowledge on insects and similarly increase their awareness of different insects in the world ,for example, rescuing the extinction experience.

The outcomes from citizen science activities provide the much- needed power required for lobbying which is important in influencing decisions made by policymakers. It has also leads to a positive feedback loop which has led to the development of strategies that are valuable, where the interrelationship between insects and people is promoted for the benefit of all (Cheesman& Key, 2007).

3.3 CITIZEN SCIENCE AND INSECT CONSERVATION.

3.3.1 History of Citizen Science

Citizen science extends to the beginnings of modern science and amateur science of the 1700s and 1800s in countries from Europe. The earliest documented projects include the Christmas bird count in 1900 by the National Audubon Society and the British Trust for ornithology that was started in 1932. Their popularity has had a steady increase through the late twentieth century and at the beginning of the 21st century and a number of such projects are still in progress in the northern part of America alone. Involving citizen scientists in projects is mostly on a volunteer basis for reasons such as data collection for studies such as changes in climate, invasive species, conservation biology, monitoring of the quality of air and water, population ecology and its restoration. Technology has also made it possible for citizen scientists to take part in studies in an efficient manner. In relation to conservation, citizens from North America gather annually to watch the migration of monarch butterflies, in a country such as Europe citizen scientists, monitor ants, cicadas and ladybugs are monitored by individuals in Asia. Another important project is that of the lost ladybug of 2011; the urban ant collector of 2012. These different projects make use of technology as a way of connecting both scientists and citizen scientists globally. The use of applications on phones such as creek watch has been used to clear some of the barriers that are met when citizen science is integrated into projects and in some cases, it has enhanced the participation of citizen scientists. The major goal of projects that uses citizen scientists is to facilitate the establishment of inventories, creation of scientific conversations between researchers and non-researchers and finally, it functions to engage and educate those participating using experiments (Hobbs, 2012).

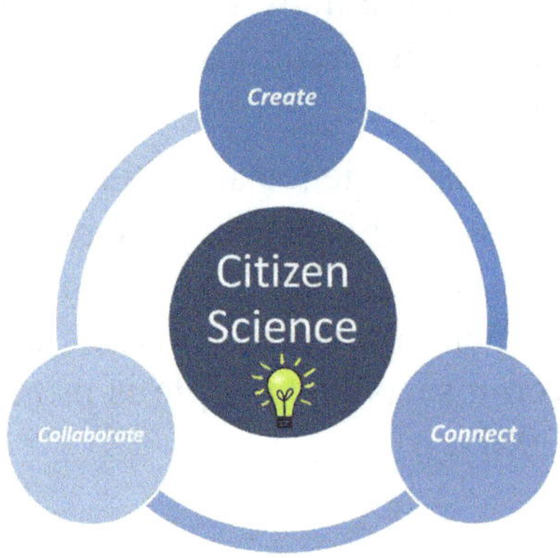

Figure 3.8: Aspects related to citizen science.

Sources https://safmc.net/citizen-science-program/citizen-science-workshop/

3.3.2 Conservation and Citizen Science

Citizen science has advanced for a number of years and has led to the development of peer -reviewed articles and have been used in making decisions in relation to the management of land and similarly, policies across different countries including the United States. Over ten years, there has been immense growth in the different states and it is ,therefore, an important tool that can be used in addressing the challenges that are encountered in conservation. They are two ways in which conservation science can be used to improve efforts that deal with conservation, the management of resources and protecting the environment. The first aspect is the development of knowledge in science while the second one involves informing policy and the encouragement of people to take action. Citizen science contributes to many fields of science and this includes conservation, natural resource and environmental science.it informs the natural management of resources, protection of the environment and policymaking but at the same time it encourages the public to be more engaged in activities that deal with conservation.

Different projects can gain from incorporating citizen science and those who use it must take precautions to ensure they match the needs for science and involvement of the public with the right type of project and the right method of participation by citizens. Citizen science is a process of scientific discovery that is rigorous and can be distinguished for conventional science other than volunteer participation. When projects incorporating citizen science are properly designed and executed provide sound science, high-quality data that is generated efficiently and it also helps in problem- solving. It is a source of empowerment individuals who are either professionals or citizens with no experience in entomology allowing them to make contributions to conservation and also research. The moth of April is the citizen science month. There are different levels of citizen science (Hobbs, 2012).

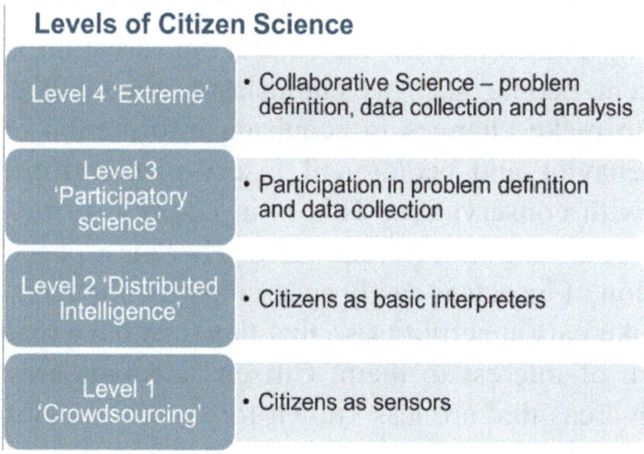

Figure 3.9: The different levels of citizen science.

Source: https://povesham.wordpress.com/2011/11/27/citizen-science-as-participatory-science/

3.3.3 Citizen Science and Insect Conservation

Citizen science which focuses on the conservation of insects is a collection of observations and facts using informed and enthusiastic sectors of the public related to the distribution and Conservation action of insect and it benefits greatly from a well-coordinated social network. In simpler terms, it allows people who are not professionals to take part in research and activities which deal with conservation but professionals are also involved in the said activities. Such research has led to the understanding of concepts in science which can be used to solve problems in the world. Citizen Science works out as a collaboration together with professional entomologists as well as programs that are well organized and established. Prioritizing insect conservation requires insect conservation from real -life decisions around awareness and access, motivation and money. Educational components are part of many programs that are well defined. Contrary to most research done by scientists, citizen science does more than just providing answers to unknown but it also goes further to combine research, education, community development and outcomes of conservation. There has to be some central point for people who believe in conservation efforts with motivation and will channel some of the efforts towards finding resources for funding in order to make conservation plans take shape. Some of the organizations that do this include bugs life and Xerces society as well as groups that specialize such as the species commission (Saunders, Janes, & O'Hanlon, 2020).

Incorporating citizen science into insect conservation activities brings together different people such as enthusiasts, aficionados and scientists who are able to make changes in acquiring information on geographical distribution, behavior and ecology of insect species for engagement of actions to do with conservation. This is in addition to the data provided by scientists specializing in different fields. There is a provision of support when the location of insects is easily accessible to volunteers and naturalists who actively take part in activities so that they can have ready contact with species that are of interest to them. Citizen scientists are also interested enough to such areas that are less known for certain species or to improve geographical records of certain groups of insects.

Citizen science means that the net of recorded data is widened greatly in both spatial and temporal scales and can feed into the development of national insect strategies. This is the case as long as there is good support for identifying the species of interest and there is an appropriate and systematic coverage of the focal geographical area. Programs that involve citizen scientists have improved the understanding of the geographical distribution of butterflies in the United States which has led to greater involvement of citizen scientists in conserving butterflies. Another citizen science conservation program led to an efficient gathering of data on the distribution of three saproxylic beetles which are charismatic and threatened, the stag beetle, the longhorn beetle and the *Rosaliaalpins* so as to improve studies related to insect conservation.it was also observed in relation to the bumblebee which was threatened in the united kingdom and pollinators in general in the united states where the volunteers became more motivated by gaining more information on bees and making contributions that are real in the conservation of bee species (Saunders, Janes, & O'Hanlon, 2020).

Figure 3.10: The stag beetle.

Source https://en.wikipedia.org/wiki/Stag_beetle

Figure 3.11: The longhorn beetle.

Source https://en.wikipedia.org/wiki/Longhorn_beetle

Including more experts across different regions means that species have not been seen for a long period and may at the same time threatened to have an opportunity to be discovered again if they still exist for example to endangered stoffberg widow butterfly in south Africa was lost in its location through inappropriate use and management of land; also as a result of operations such as mining and pollution.

Africa is one of the continents with the most diverse species of animals and citizen science has a role that is important in insect conservation ,for example, the limbovane outreach project in South Africa, a project that has managed to successfully engage both teachers and learners in rural areas to recognize the value of ants and assist in the conservation of the megadiverse greater cape floristic region. Prioritizing the conservation of insect species depends on different aspects such as the size and their brightness of color, the level to which an insect species is at risk of extinction, whether it is on the red list; how much people care about it; its importance and whether people are willing to put efforts and money into it for the purpose of conservation. From a global point of view, this approach is considered as a means that is biased but it is important for species that become lucky enough to get attention. This also benefits other insects and biodiversity as a whole as it creates focus on the problems insects deal with. Citizen scientists play a role that is important in seeing the actions related to conservation take place. In the United Kingdom, the white- faced dragonfly is threatened. This only occurs in some sites with acidic bog pools with floating sphagnum moss. In the country, a site known as foulshaw moss was identified as an area for the species from one of the three remaining localities where the species exist. Collaboration between the British dragonfly society and Cumbria wildlife trust led to the reintroduction of species between 2010 and 2o14. Their results were fruitful and by the end of 2016, about 2000 exuviae were counted. While the landscape approach aims at increasing heterogeneity and functional connectivity on the landscape for insect and others biodiversity, at species level there may be the need to focus on specific habitats which would include not only the right biotic variables such as a host plant availability but also abiotic variables like microclimate. This does not mean that the habitat or locality is viewed in isolation, but rather than that it is seen as part of the larger and resilient landscape (Colla, 2017).

3.3.4 Research and Conservation Science

Not all research on insect ecology can be of benefit to citizen scientists but it is a means that is of value in responding to queries related to the ecology. Relevant projects include those that have large time related and geographical measures are of value. At least fascinating parts of the research and which nonscientists can have entry to can be easy to the critic. It is made simple if those putting these programs in order have an appeal to and interest in ecology. When researchers use citizen scientists, they need to defend their quality of data due to the presence of imprecision that may occur such as observation inaccuracies and inaccuracies that result from observations that are nonrandom and they should develop ways of avoiding them. Accurate strategies for data collection can be developed by providing training sessions that are person to person and also practical coaching. The progress of activities related to insect science should also be monitored and evaluated to determine the effectiveness of such programs or projects. Some of the projects may have a kind of simplicity in which it is easy to notice the results or outcomes. The interests of active participants can be sustained through the use of newsletters, social interactions and finally the use of meetings (Colla, 2017).

3.4 INCREASING PUBLIC PARTICIPATION AND ENGAGEMENT IN INSECT CONSERVATION

Engagement of citizens in insect conservation can be achieved using different means, this includes social- political action in relation to local issues which deal with the environment; projects to do with conservation, stewardship and monitoring; involving them in scientific investigations and the use of education and outreach activities. These can be done individually based on the type of project but they are mutually exclusive. Members of the public who participate in research are referred to as citizen scientists and on the other hand, those who take part in other conservation activities such as restoring insect habitats or those doing advocacy work may sometimes not be referred to as citizen scientists since they do not assist in the collection of data, making observations and therefore they are do not take part in scientific processes. Although this is true, both efforts are necessary in order to achieve the desired outcome towards achieving the conservation of insects.

3.4.1 Education and Publicity

Education provides a platform to increase people's awareness of biodiversity and its needs. There is a need for enhancing awareness. In areas ,for example, a study done on awareness in Australia noted that

- There was a lack of awareness widely especially on the importance of biodiversity and process in the ecosystem.
- Citizens are a driving force to propel the concept of biodiversity in different regions
- People from urban areas usually have more knowledge and are likely to support the conservation of biodiversity.

A lot of observation has been made in studies, that there are a decrease and loss of interest in natural history especially among the young generation and there is a lack of attachment of people and their natural environment as a result of them adopting a sedentary lifestyle especially due to technology and urbanization and thus the common phrase extinction of experience. This loss has been caused by loss of opportunity by people moving to urban areas which are separated from natural ecosystems most of the time. Another cause is as a result of losing orientation due to the reduction of emotional sympathy with nature since they lack connection with nature because they lacked it during their childhood. This ,in turn, changes the attitudes and emotional responses which give rise to a loop of feedback associated with a lack of affection to nature. It is commonly observed in urban areas and numerous educative and incentives ideal with insects in an attempt to stimulate their interest. Approaches used include the creation of children illustration books with insects in ways that increase their interests in insects, development of homes that house insects to advocate for their study and rearing; development of museums to stir interest and educate people on the biodiversity of insects. Creation of publicity through different media channels by the governments to share knowledge on natural history and also increase awareness of needs related to conservation and how they can be achieved (Hobbs, 2012).

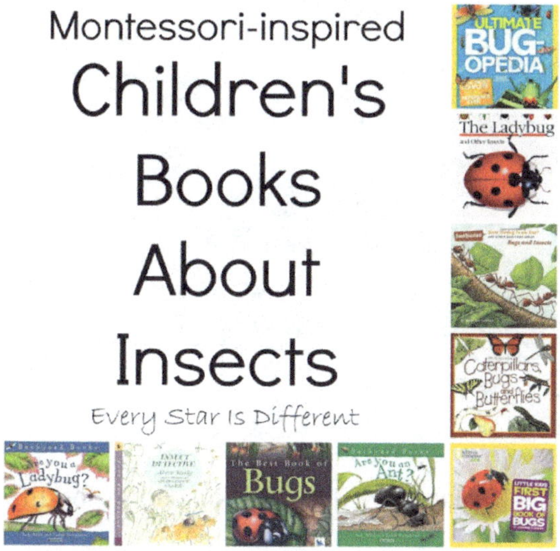

Figure 3.12: A children's illustration book.

Source: https://www.everystarisdifferent.com/2016/07/montessori-inspired-childrens-books.html

Outdoor learning can be used in formal education can be important. Formal education can be used to provide opportunities for early development for example skill identification. Studies have shown that a memorable setting can be used to enhance learning for example field trips. Field activities go on to create more positive attitudes and it also allows people to develop socially. It has also been reported as one of the activities which bring enjoyment to students and can be rewarding to them. It has been identified as a way of motivating students to pursue subjects at higher levels of education and also making them adopt behavior that is friendly to the environment. The concern is that there is a decline in opportunities for outdoor learning in schools and higher learning institutions. A study done in the United Kingdom found that a small percentage of schools that are less than 5 % provided learners ages eleven to sixteen with a platform for field courses. There is also a decline in fieldwork quality and quantity for students in schools who are above sixteen years. Field activities should be encouraged at all levels of education allowing students to connect with nature and this will create and show the need to conserve and protect both local and global biodiversity (Hobbs, 2012).

Figure 3.13: Students planting milkweed to support the life of monarch butterflies in an outdoor activity.

Source: https://blog.education.nationalgeographic.org/2020/01/14/educator-spotlight-supporting-monarch-butterflies/

Figure 3.14: Students sharing their knowledge of monarch butterflies with their school community.

Source: https://blog.education.nationalgeographic.org/2020/01/14/educator-spotlight-supporting-monarch-butterflies/

3.4.2 Short term Public Events

Citizen science projects can last up to a number of days and may include the counting of insects or the participation in surveys by the public. Some activities include butterfly counts, BioBlitzes festivals and national moth week. It is an event that facilitates the education of the public but does not make any contribution whatsoever to the monitoring of collecting data related to science.

One-day butterfly count begins in the North of America as the 4th of July butterfly counts which was started by the Xerces society in 1974 and the inspiration to begin this was gotten from the Christmas bird counts which were done by the Audubon Society and were started by Frank Chapman in1900. The North American Butterfly Association collects data from doing these counts and it has been noted that the numbers continue to decline. This is related to the aspect of phenology which is a barrier to making comparisons of butterfly populations that are counted. One of the main reasons for participating in the activity is to measure the abundance of butterfly populations. Most of the time it measures the peak emergence of a species in a single year and it is compared to the beginning or end of the emergence of the same in a different year. The decline may be due to the fact that some of the counts are not availed on online platforms NABA does not keep sightings that incidentally occur on their website. Another reason for performing the counts is to increase appreciation of butterflies, their diversity and the services they offer to the ecosystem. The data over time can provide different insights such as an extension in ranges and can take not of numbers which are not usual (New, 2012).

BioBlitzes involves detecting and identifying insects and similarly other species. They are usually taking place during a single day or once and some can even take days. Experts take part in the activities but are helped by volunteers from the public. The role played is to survey a certain species found in an area or a snapshot. The term snapshot considers phenology. An example is BioBlitzes Canada noting all the species that exist and also the discovery of others. Such activities can be used to create awareness on biodiversity. Like any natural history outing, they have the potential for making interesting findings in relation to new species. The most recent event by BioBlitzes in the city nature challenge in which people made the attempt to find, identify and record living things in their locality. The partners working with bio blitz also made arrangements for events in national parks to create a connection between people and nature (New, 2012).

Figure 3.15: Connecting people to nature a bug hunt in Hyde Park which took place during the city nature challenge in London ,2019.

Source: https://pentoprint.org/thursday-connectors-senses-of-nature/

Festivals with insect themes are organized to appreciate diversity and also promote awareness. Such events occur in nature and science centers. In Alberta, the Ellis bird farm, the bug jamboree is done yearly and participants are both amateurs and professionals with presentations and exhibits. An insect theme song is sung to mark the beginning of the event and to end it a butterfly count is done. Such festivals are events that provide education more than events that are used for collecting information.

Another example is the national moth week is an attempt making use of online platforms with the aim of coordination and encouragement of events to appreciate moths and it engages people worldwide. America registers 245 events while Canada does around fifteen. One of the major objectives is to produce a database on moths globally and similarly to most events that take a short period of time, it can be said with clarity that education is the primary function and also to the promotion of their conservation. Another program is the school malaise trap program in which students are provided with malaise traps to use in their school fields for free and the specimen they collect is center for biodiversity genomics at Guelph University and in return, they are provided with reports on their findings.

3.4.3 Online Database Projects

They are projects on databases that the public has access to and it can be used to in the characterization of the essence of science that is modern. Atlases are precursors of such projects and similarly to butterflies count their inspiration comes from projects related to birds with a similar nature. The most common species used in atlasing projects are butterflies for example in countries like Canada. The area that is subject to atlasing is overlain by a map grid and its division is done with the use of squares and later a survey is done. Naturally, it requires different visits to the same square due to changes in phenology which occur in the assembly of butterflies for a given period. For traditional examples, the end result of or atlases is books, if they are new projects the information is made available online on their website. Projects related to atlasing happen over a determined period before the announcement of the complete result but they are becoming increasingly open- ended with the use of maps and other summaries of data and this information is updated continuously online. Butterfly atlases are different from other butterfly database projects which are open- ended this is because they encourage those participating to visit the different squares on grids used in atlasing. Many insects are not involved in atlasing because it requires a large number of people to participate in areas with large geographic coverage. Some of the projects include species such as Diptera in the United Kingdom and also the Odonata. One very successful project is the butterfly databasing project in North America and it is referred to as eButterfly which bases its inspiration from the eBird project which was organized by the Cornell laboratory of ornithology.it has been used to make contributions to iNaturalist Canada which are controlled by the Canadian wildlife federation. It refers to itself as a community of people who are interested in nature. And is under control of the California academy of sciences and currently, it has more than five million records some of them are based on photographs and can be considered as research grade. Many citizen science projects share their information iNaturalist which goes on to share its data with the global biodiversity information facility.

ebutterfly was a concept that was introduced by maxim larivae who was a doctorate student together with Jeremy Kerr at the University of Ottawa. The expansion of this project has even reached the United States. Experts volunteer in each state to check incoming observation and this may have photographs present or not, field notes and sketches.in comparison to eBird, eButterfly has a scope that is much smaller and has received support from the team such as sharing knowledge as well as information on their successes and their shortcomings. The emerging them from their dialogue has been the significance of making the database useful to people involved in the conservation of nature for their own purposes. eButterfly is motivated by different concerns. One of their primary approaches is the checklist approach in which those observing provide information on the butterflies as they see during an outing, and they ,therefore, provide data on the presence or presumed absence of some kind that is more helpful in the analysis of the ecology than records which are incidental of individual butterflies which are worthy of taking note of. Butterfly projects divide and group citizen scientists into those that are motivated as naturalists and those who are concerned with conserving nature. They mainly focus on one species for example the Pieris project mostly wants its participants to send specimens of cabbage white butterflies to one person who coordinates the activities which are used in a significant project. A larger team of researchers use this data to look into patterns globally in the differences with DNA sequences in species that have been introduced widely.

The monarch butterfly, is one of the species examined in citizen science because of its unique nature and there is a concern about its annual migration and it is sometimes seen as a flagship species. One project concerning them is the monarch watch and it entails tagging them in the United States and Canada with the hopes that they will recover on winter grounds.

Other species that inspire projects are the Odonata such as damselflies and dragonflies and have a website dedicated to them such as the Odonata central. The website contains checklists which are regional, distribution maps of the species and photography on the insects. Another project is the dragonfly inventory and it is overseen by Paul Brunelle and has support from non-scientists.

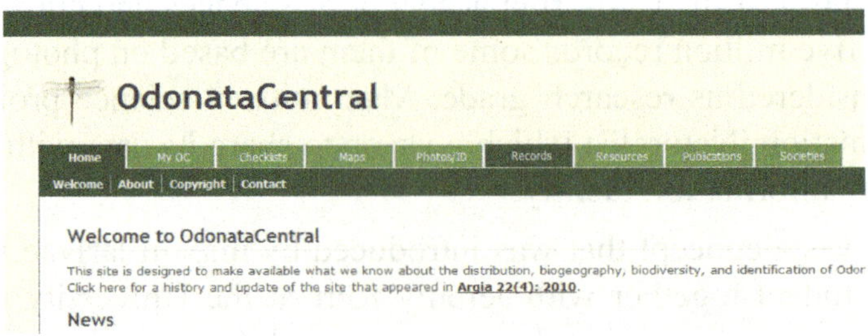

Figure 3.16: The Odonata central website.

Source: https://thedragonflywoman.com/2011/09/23/

Another development in citizen science in the north of America is the bug guide https://bugguide.net/node/view/15740 which has information on insects and other arthropods. Photographs are checked for validity by scientists on a voluntary basis. The arrangement of the arthropods is according to their current taxonomy and this is updated on a continuous basis. It also provides information on natural history and hosts a number of forums with relation to it and it facilitates discussions on identification, taxonomy and photography. People also gather in different locations in the United States and it is ,therefore, one of the most used websites for looking for information on insects.

Figure 3.17: The bug guide website.

Source: https://www.ecobeneficial.com/great_resources/bugguide-net/

Other sites also put their focus on bees and other pollinators and concerns on species that are introduced. One of the most popular projects is on bumblebees that is the bumblebee watch. It requests for photographs on bumblebees which are then identified by insects and the data from these is added to a database and plotting is done on maps. The monitoring of the species is done and is due to their population decline and brings different people on board.

Figure 3.18: A citizen science program; a child collecting bumblebees for a survey and volunteers taking photographs of a reference collection of bumblebees.

Source: https://wildlifepreservation.ca/blog/bumble-bee-watch-using-citizen-science-for-conservation/

One recent example is the early detection and distribution mapping system which monitors newly introduced species in the United States. This means that citizen scientists are asked to give reports on invading species in their locality and this information is availed to authorities who are required to exercise control measures as seen fit. From an economical perspective, it is important to develop such a database from a standpoint of conserving biodiversity. In the 90s, Canada set up the Canadian Nature Federation started a project referred to as the ladybug survey which was done both in schools and outside using magazines. This project led to 35,579 records which were made available on their website. Errors in identification were a recurring issue with the dataset and maps and this led to it being brought down. The project focused on the concept that monitoring an invader would be a tool in controlling it and as seen from the results it was noted that there was no reduction in the spread of the coleopterans in Canada. The lost ladybug projected is found in the united states and it was fruitful in monitoring species that are native or introduced and those coordinating it were able to learn lessons of value from the experience of their predecessors, which focus on the thrill of discovering species that were considered lost and were now not common after the introduction of another species (Didham et al, 2020).

3.4.4 Clubs, Societies and Social Media Groups

Citizen science also includes the participation of amateurs in societies and clubs related to the study of insects. This encourages their participation but can either be successful and others can sometimes be unsuccessful. Many amateurs are interested in taxonomy and fauna in comparison to other complex fields. Some of them possess the training to understand some of the statistical analyses related to entomology. Other areas of interest include ecology and behavior. Scientific societies that are large and meet annually, make it easy for those interested to connect with their interests and the disincentive may be access and also the costs they pay as members thus more amateurs will be interested in societies that are small and local due to their affordability.

There are several examples in Canada include the Toronto Entomologists Association and the Alberta Lepidopterists' Guild. The later incorporates citizen science and can combine group projects and the use of social media. In one of its projects the encouraged people to take part in eButterfly. The Alberta Roundup was started in 2015 and it encouraged those participating to re-confirm the entire species of butterflies found in Alberta and it is still continuing. There were one hundred and twenty two species that were confirmed and in 2016 another number was added. The project fosters competition that is friendly to serve as motivation for the group and acts as a reminder to those participating that a species is on a checklist doesn't mean that their existence in an area in continuous. The roundup also encouraged participation in a number of ways such as submission of records using an email list server, Facebook and also in person.

Another example of using social media to encourage participation in groups was the Alberta winter bug count which was not sponsored by a specific organization. The primary purpose of the project was the comparison between insect and bird diversity in Alberta during different periods. The records were from three different list servers but still preferred presenting the specimens directly to a central individual and it is important that those compiling allow people to use comfortable means and the fact that not all people are tech- savvy (Didham et al, 2020).

3.4.5 Crowdsourcing

Another approach to citizen science is crowdsourcing and it makes use of volunteers to transcribe specimen data labels. An example is the Notes from Nature project through the Strickland entomological museum at Alberta University like the Smithsonian Transcription Center. To make collected data digitized, photos of the specimen are taken next to their data labels and these are made available online at an activity that can be fun. Interpreting labels that are handwritten can be a challenge similar to identifying locations that are obscure especially if it contains a name that is local with a broader geographical context.

Another example of how crowdsourcing can be used is the use of volunteers to help in the alignment of data sequences related to DNA in the project Phylo. It involved people from the United States and Canada since the human brain is better than using algorithms in the recognition of long patterns such as DNA sequences. It recognizes the fact that the participants don't need to have awareness of the matters. For different individuals, the activities seem to be simply busy work and the motivation is the thought that it would lead to results that are of benefit to science. In some cases, the citizen scientists do not get to learn much about history and natural science and therefore will be dependent on the type of project.

4

The Conservation of Insects By Protecting and Creating Their Habitats

CONTENTS

4.1 INTRODUCTION

Insects exist everywhere you look and there is a wildly diverse number of Species that's it under the class Insecta with scientific studies estimating a total number of nearly 9 million prescribed but many estimates there could be up to 30 million still yet to be scientifically taxonomist. They are the planet's most varied collection of creatures and it is estimated that 70% (some entomologists say maybe as much as 90%) of the different species of Earth belong to the class of insects. This makes them probably the most successful creatures to have ever existed on the planet and this can be attributed to the tiny size making their energy requirements easy to manage and their wild diet makes this challenge easier still. They have a tough exoskeleton that protects them and many possess wings that allowed them two find food which safety and possibly mate fairly easily and they have a prodigious ability to reproduce.

Interestingly enough to have been able to cover every surface of the planet except probably Antarctica and they are mainly landlubbers and with 70% of the earth's surface is covered by the ocean it is the only area they have been unable as a species to penetrate. There are however some fairly aquatic insects but they do not inhabit the largest biosphere on the planet. The physiology and morphology together with a vast range of habitats have seen them survive beyond and in fact, the oldest insect fossil is somewhere in the range of 400 million years old which probably suggests evolutionarily they were one of the first animals to translation from living in the sea to becoming terrestrial animals. This means they actually were around a good 170 million years before dinosaurs even came into existence.

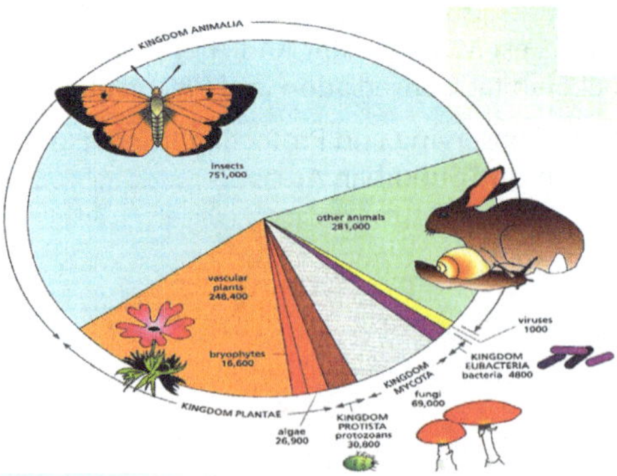

Figure 4.1: 70% of the species on Earth are insects, despite the land covering only a third of the planet. Insects in terrestrial and freshwater systems are the most highly ecologically connected of all organisms.

Source: https://www.slideshare.net/SmitaShukla1/introduction-to-life

Human beings have a great capacity to think well beyond their importance believing heavily in the omnipotence however the why do you very ecosystems and biomes that exist everywhere that even human beings do are dependents on other living organisms or entities that make up what is ordinarily studied in science as the food chain. Most if not all ecosystems and biosphere are organized in such a way that requires the interaction between all these organisms to take place for a healthy ecosystem to exist and insects are a vital structure performing many important functions.

Insects are most in this aspect as they pollinate plants in for example how honey bees are able to and they serve as a food source for many birds, mammals and fish. Insect plant relationship is probably the most important biotic interactions on the planet and even though in terms of biomass plants probably exceed several times more all animals put together you will find that there is a functional significance that insects play. In this way we can say that there is a lack of human appreciating for the importance of insects and even physics officially there is a disregard and sometimes is dislike towards insects but they do serve human beings positively directly and indirectly. Honey bees for example produce honey about 1 million tons every year and also beeswax. The larvae of silkworm moths make silk from their salivary glands which is the essence of an industry of its own that is valued at an estimated 1.4 billion dollars. Most insects like these serve a direct function either through what they produce or in controlling agriculture. Others are a food source too many birds, mammals and fish which are a direct consequence of the fishing and hunting industries. Ecologically within the food chain, they help decompose dead material ,which reintroduces vital nutrients back into the soil and these can be observed in the chapter discussing soil insects' habitats to be highly beneficial to plants even with burrowing insects that provide water channels.

For human beings, insects can also be a source of food and are high in protein for example in Mexico and many parts of Africa where they are sold in the market and are mainly fried before being consumed. By weight insects or actually, a much better source of protein than the conventional beef chicken purple lamp and they are also much lower in cholesterol and fat.

Figure 4.2: Deadwood, like many others seemingly unlikely but very natural micro ecologies, are extremely important habitats for a variety of insects acting as their own micro ecologies.

Source: https://www.amentsoc.org/insects/conservation/habitat.html

4.2 INSECT SURVIVAL AND HUMAN IMPACT

During the last 1 million years give or take it can be geologically noted that glaciers have advanced and retreated some within even a few decades. Regardless, insects have actually survived and overcome even some of the most catastrophic extinction events with the shifting back and forth of cold and warm fronts over the last 6000 years or so however human beings began extensively manipulating the environment in the landscape around them first destroying most of the extensive rainforest even as far north as Europe and changing the various hydrology systems. These actions may have definitely lead to some local extinctions like the Neolithic habitat for beetles around Britain that died of in areas that are now devoid of trees even up to today their Extinction rates has been called as the great insect dying on a global scale calling it the sixth extinction event on Earth. Some scientists extrapolation based estimate that suggests over 11000 species of insects have gone extinct since 1600 and over the next two centuries some state that many as half a million insect species could actually go extinct if human endeavors and development do not change course (Kumar, Yogi, &Jagdish, 2013).

Despite their importance, it is only just recently that many organizations have begun mainstreaming activities that go towards biodiversity conservation. There are many global insect conservation initiatives that have produced solutions for Humanity on how to conserve insects realizing that our fate is intertwined especially through the medium of plants. Like the British Bumblebee Conservation Trust and the Royal Entomological Society.

Some scientists and analysts have found that more than 40% of insect species are declining and many within that are endangered with the rate of extinction being 8 times faster than mammals Birds or reptiles. With huge losses already been reported in large animals, it is not a far-fetched notion. The biggest cause of this insect decline and the highest threat has actually been in the way human beings produce their food and the repercussions that they have on the planet's ecosystems and it is catastrophic, to say the least. The overuse of agrochemicals such as insecticides herbicides and fungicides has hit hard on the insect world. Fungicides, for example, which are actually harmless directly to insects can interact with pathogens to actually end up harming insects while insecticides and pesticides are intentionally built to destroy them. Overpopulation is also played a huge Factor in insect decline with having many habitat areas cleared either for human beings to live in or for human beings to produce their food and currently it is argued that the Earth is 60 to 70% over capacity and give the population of human beings on earth reaches 10 billion we are arguably going to need another earth hypothetically speaking. Because of this and other activities insects are said to be vanishing from the planet at a rate of 1 to 2.5% annually but this is under debate in the scientific community with no verified evidence making this statistic highly speculative (Kumar, Yogi, &Jagdish, 2013).

Figure 4.3: Icariciaicarioidesfenderi, sometimes known as the Fender's blue, is an endangered species in the United States. In the 1990s, the species numbered less than 1,500, but managed recovery plans have raised its numbers to as high as 25,000-30,000. They did this by conducting a prescribed burn (an intentionally set wildfire) in its habitat.

Source: https://entomologytoday.org/2019/08/16/look-science-insect-conservation/

One place insect decline was studied quite extensively is in Germany where over the last 27 years it was seen that insect populations decreased by 75% in natural areas. This study prompted the Dutch to react in very critical areas with the Delta plan for biodiversity recovery. The Delta plan for biodiversity recovery was a declaration signed by a strangely unique collection of data scientists, farmers, nature and environmental organizations, participants in the food supply chains and a bank as well in 1917 to create the Netherlands where people and nature flourished side by side. Direct actions included The Creation management and maintenance of green spaces by farmers, Dutch supermarkets were also incentivized to create a new set of standards and quality labels and around 25 administrative units of the Dutch Society for the conservation of nature to create initiatives work hand-in-hand with farmers in order to meet certain objectives like creating conditions that are beneficial to insects where they may be able to flourish (Samways, 2019).

So insects ,as we have seen and as the Dutch and many other countries, have made conscious efforts for are worth conserving in that we share the same planet with them and broadly put we need them for our own survival and our economic dependence on them is quite high and naturally as human beings, we should recognize the moral responsibility to act as stewards of nature showing respect to all other life-forms.

4.3 THE CONSERVATION OF INSECTS

Probably the most fundamental insect conservation tactic is to protect and where necessary or applicable and hand and even create habitats throughout the landscape so that even the least mobile species can move around and find a place to live where the local conditions have turned against them. We will talk about habitat conservation more specifically in the rest of this chapter but it is important to note that even if habitats are conserved successfully insects can be harmed by pesticides and many other different forms of pollution. Like stated above pesticides are widely used across the globe and have been one of the major causes of insect decline over the last several decades and do they sometimes are the only economically practicable way of controlling a small minority of agricultural pest they do a lot of damage to non-targeted species many of which can actually serve to be beneficial to farmers and growers (Collins, 2012).

The use of pesticides it can be reduced in they are harmful effects by using farming methods such as the integrated control which is where a variety of different Tactics are combined so that we lower use of pesticide and sometimes even eliminating the dosage overall. Pesticides have actually been proven to be doing more harm than good in society as many people with autoimmunity issues have health effects derivative from their diets which they consume plant material that has been affected by pesticides sometimes consuming these pesticides directly and excluding them from farming by using the integrated method will actually have further effect even beyond the conservation of insects. Where pesticides have been applied like there are some places where there is no alternative they should be kept away from Hedgerows water systems or watercourses and basically any other area that can affect the non-target species. It has actually been proven that one of the greatest problems when it comes to the conservation of an insect over the last few years has been the contamination of watercourses with sheep dips that use or are based on synthetic man- made pyrethroids. A pyrethroid is an organic compound that can often be found produced by the flowers of pyrethrums that is sort of a natural insecticide and is quite similar to natural pyrethrums and is used in the majority of household and commercially sold insecticides. They are harmful to virtually all invertebrate life and have proved to be very important to such an extent that it can illuminate all insect life even miles down the stream from where the smallest of leachates occur (Collins, 2012).

It can be viewed that insects are at less of a threat let's say various vertebrate species that are on the decline due to the fact that first of insects exist in the far larger numbers than vertebrates and secondly they reproduce in far larger numbers as well although many of them will die due to natural causes and hazards. For some though the hobby of collecting insects may be a good idea in their conservation if the existence exceptionally small: this can pose a potential problem in that they can easily be killed. Sometimes especially when large numbers of these insects are being traded by collectors, there may exist legal restrictions as well.

The usual course is to have authorities who conduct a scientific assessment before the place any species on a protected list so that to avoid any unnecessary restrictions and regulations that can further discourage people from having a sincere interest and studying these insects adding to the knowledge base that already exists that is important towards conservation efforts. Obviously, this situation differs from country to country and needs to be observed wherever you live.

Figure 4.4: One of the greatest tasks facing insect conservation is overcoming the perception challenge where most people find many insects as repugnant or a nuisance and unworthy subjects for conservation action.

Source: https://www.researchgate.net/publication/247844038_Insect_Conservation_Biology

4.4 HABITAT CONSERVATION

In areas where insects used to initially flourish the numbers and those of various other invertebrates is on the decline and the fact that human beings are increasingly destroying their habitat is probably the primary reason why we are human beings need to be concerned about this issue with the main solution ,therefore, is conserving habitats. Habitats are areas where specific environmental or ecological conditions exist in which an organism lives and reproduces. Many animal and plant species have very specific requirements towards this respect meaning that these organisms will only be found in the area where requirements are met. Conservation is the term used to describe the maintenance and protection of biodiversity which is far-reaching and may include anything from land management techniques to education and raising awareness on how to behave towards life-forms. In this chapter, the focus will be on the conserving of a particular habitat that is important to the species in question and that is insects, and it may be a lot easier said than done especially since the destruction of their habitat is usually in relation to our search or development of food housing and overall prosperity. Leaving anenvironmental habitat for insects to inhabit and living is subsequently a price that needs to be paid by the adjacent society which means ,for example, reduced agricultural yield or maybe extra costs for diverting means of transports like roads and railways around a protected wildlife site and many others along that line. The question here however is one of perspective because if we decided to look the other way which is on the other hand failing to conserve their habitats we have even a greater price to pay as insects are so important to our own very existence (Collins, 2012).

One may be quick to ask, if insects have been so successful perfectly well without human intervention for as long as 400 million years why would there be a need to look towards conserving their habitats now? And the reason is even though our ancient ancestors still destroyed habitat so that they could fulfill their quest for all sorts of food safety and other priorities the real problem here is that we are now so numerous on Earth and we destroy the habitats on a much larger scale especially with modern technology and was previously possible and this makes conservation so vital. A question may also arise as to the conservation efforts that have already been made where nature reserves have been set up to protect wildlife and it may be tempting to ask whether this is enough and the real answer is it isn't. It is actually not enough with the current level that exists all across the world to have the current nature reserves especially if you want to conserve all the species that currently exist. The main driving Force towards setting up a reserve is based on the reasoning that we need almost all of the Earth's resources that actually exist right now for human purposes and so we set aside some small section of these areas as wildlife reserves and in theory as much as they could represent a series of biomes in all different kinds across a particular region they will not be able to actually support all the species of the region in question and so it is important to continue with habitat conservation effort beyond what we already have since it is also impractical to set up a full range of representative biomes.

To clarify whether it's also important to recognize that the habitat does not actually last as unchanged environments for extended periods even in the same locality and so a Noah's Ark mentality will fail. From one year to the next there are different weather and climatic fluctuations that could end up creating situations where the insect in question may not be able to find everything it requires and in some cases produce unsuitable situations even temporarily that could change the habitat against their favor. If thinking in this particular line then it is practicable to have a large reserve and maybe even extensively so with a number of smaller suitable localities that exist near each other to allow a greater probability of long-term sustainability. This will require proper management techniques because even this large reserve can lose species if the diversity is not dealt with appropriately. This would mean that they're the most successful way to initiate this idea of conservation of habitats is by not only having protected areas or nature reserves but on top of that create almost in a sense our own protected smaller areas consciously where we can even within our everyday landscapes whether they may be urban or rural forests or towns, etc. From an entomological point of view, this is known as creating Meta populations so that you have a network of very dynamic insect groups of particular species (Collins, 2012).

To a certain extent, we can manage we already established protected areas so that we have better habitats to provide for the insects but da issue is no amount of good management will actually entirely remove the need to have conservation habitats throughout the landscape. This is because we may be able to mentally envision the habitat requirements of a particular insect species or even create and maintain conditions for that species which is known as gardening the habitat but the problem is it will only benefit may be just that one species or at best a few species which will end up by extent harming many others or even driving them to Extinction on the local ecological system.

When we look at conservation of insect habitats we have already recognized that habitats in themselves are being damaged all over the world and it requires a multifaceted approach for the different situations to be addressed. In places for example where there is an extensive natural ecosystem like tropical rainforests especially on a large abundance scale, we must find it our imperative Duty to leave it as intact as we possibly can but this is far-fetched. Most Tropical rainforests which actually host more than half of the species of insects that in the more economically challenged countries where policies are difficult to enact and implement; so help from the more economically prosperous countries may be needed to slow down or entirely halt the rate of destruction.

Beyond that even though they are probably many insect species that exist which depend on Natural ecosystems that are many others that share our habitat do we already mentioned on the land that we use for the production of crops and in and around our houses anywhere outside and even some in industrial sites. List of below some of the insect happy shops and the breakdown of house American be taken to conserve these habitats for insects and other and vertebrates (Cheesman, & Key, 2007).

4.4.1 General Farmland

- Leave hedgerows intact and do not destroy them. Dry stone walls are also important insect habitat and should be left pristine to amalgamate fields.
- A hedge should not be cut on both sides in the same year. This gives you a better long-term chance of survival for very many in vertebrates that inhabit arable fields.

4.4.2 Arable Land

- The use of pesticides insecticide and fertilizer should be avoided at all costs and where this is not possible it should be minimized especially around the edges of fields and again near hedgerows.
- Where arable fields exist and they are very large one should create uncultivated strips also known as beetle back across them so as to keep the natural habitat as natural as possible and also to encourage biological enemies of pest species.
- It is important to mention time on again how discriminate see her besides in pesticides should be used and also to a donut maximize

the Habitat potential by introducing the strategy of set aside land and providing a good range of insect food plants.

4.4.3 Grasslands

- Unimproved or natural grasslands should be retained
- In these areas ,woody plant should be allowed to grow especially where it is necessary to control the scrub so that they can accommodate some species of insects
- Herb rich vegetation is also a great option to restore in grasslands which have to be reseeded
- Where grasslands are used for grazing cattle it is helpful for insect conservation to have a controlled raising technique to create a combination of different heights in the expanses of grass. This helps avoid both ove grazing and under grazing. The traditional breeds of livestock native to the area will do this naturally and are suited for local conditions to help this kind of management.
- Again the use of an artificial fertilizer will most likely damage the local insect habitat and where possible and economically viable the land should be converted into herb- rich conditions if you want it improved.
- Pasture woodland trees should be protected as much as possible because the areas around the base of the trees are usually free from trampling livestock and serve as great insect habitats
- Ivermectin and other veterinary drug are damaging to insects that feed on dung and should be avoided at all costs.

4.4.4 General Woodland Area in Deadwood Habitats

- A good portion of the trees around these areas should be left to age, die and decay naturally instead of attempting to harvest all of them together.
- The edges of Woodlands would be kept open and sunny maybe edged by a few shrubs
- Ancient trees should be protected which many exist in woodlands rather than thick forests to allow the younger trees to subsequently take their place over time
- Leaving firewood in piles is a bad idea as they may attract rare

local deadwood invertebrates that make these materials their home.

- Avoid planting in areas where wetland or grassland microhabitats exist especially where woodland schemes are being set up
- Regular coppicing can also be helpful but should be avoided in areas where the local insects depend on sap runs or even Deadwood. Here, nature should take its course

4.4.5 Moorland Areas and Heath Areas

- A heath is basically a shrubland habitat built up of acidic soil making it virtually infertile. Low growing and quite woody vegetation is seen and these areas and it is normally related to the high ground close to Mountainous Regions. Again here a portion of trees should left to age and decay but periodically cutting and sometimes even burning small areas is acceptable but burning of large areas is not.
- The cutting down of some trees in the area can be helpful so as to protect the scrubs and the open Habitat but it is important to leave some formal shelter for Habitat diversity in the area.
- Pictures of strips of the bare open ground are also important throughout the area for bare ground Habitat insects and this is why some form of controlled burning is acceptable

4.4.6 Aquatic and Water Insect Habitats

- Deepening on the straightening of streams or rivers should be avoided as aquatic insects do not actually live on the water but depend on shallow edges.
- Where do your situations a video card schemes to reverse such changes a possible by restoring gentle shelving edges?
- In these areas, open vegetation is important such as reed beds and marshy areas and some woody plants if they grow there should be left alone because they are unique natural habitats for local species.
- Woody debris around the area like falling timbers should be left alone unless they pose a significant risk to the people of property around the area due to potential flooding

- Silted areas like ponds and ditches a good not only for the local fauna but also for the flora.
- Any development schemes in flood Plains should be avoided as they could involve destroying the habitat.

4.4.7 Parks and Gardens

- The local park and gardens should provide some wild areas even in the micro scheme and where formal and tidy appearances are required they should exist next to the paths by simply maintaining need lawns and flowers.
- Garden wastes should be kept in the compost heap for future use
- The plants that are grown in these areas should be those that provide a habitat for the plants that are grown in these areas should be those that provide habitat or general use for a wide number of insects rather than just nectar sources for example
- If room allows, there should be dead with Habitat and water Habitat also created in the same area to create diversity. A tree that has been cut down for safety reasons for example should keep a tall stump.

4.4.8 Brownfield Site

- Brownfield sites are essentially abandoned industrial areas and some can be suitable for the creation of microhabitats for invertebrates.
- Brownfields can host a biodiversity range like those of ancient woodlands, so education should be pushed towards having to redevelopment as the only option as they can also be turned into greenfield lands and protected habitats. If redevelopment is the option of choice then plans should ensure that the Micro Habitat in the area can be retained and properly manage afterward
- Reducing the current insect decline relies on a philosophy of appreciating that human and insect survival depends on mutual well-being while noting the obvious difference between the financial gain in terms of service provision and conservation value in terms of the irreplaceable nature that insects have in maintain the ecological integrity. We then need to understand the current dangers facing insects, and aim to maintain their various

habitats. To do this, we need to realize a deep appreciation of the spatial scale, most easily propagated through conservation of the landscape.

4.4.9 Tropical Forest

The conservation and protection of this insect and general fauna habitat has many undertones in the complex working of the global ecological mechanisms. Tropical Forests including its various epiphytes host a wide variety of insects providing them with a warm and moist environment consistently. It is rich in species from quite a number of taxonomic groups. What this means is that with the meaning species comes a myriad of ecological interactions some of which are highly technical both in the canopy above the ground and own the ground involving the delicate soil. Maintenance of these areas require large scale conservation efforts to keep the forest area intact which unfortunately is rapidly declining. The idea here is to have specific preservation goals which fortunately for insect habitat conservation and protection specifically is not very difficult because entire forest patches with seemingly no conservation status have insect conservation value to a certain extent and this happens more especially where you have high-quality fragments that need to be protected but are not exactly connected through these patches. This is where we begin to form a sense of functionality and eventually, special attention is given to them and the same case applies with buffer areas developing meaning and conservation importance in their own right (Samways, 2019).

This is not enough however because areas that have undergone degradation over the last couple of years or maybe the last few decades need to be recovered so as to regenerate and this takes quite a number of years especially when thinking about the recolonization of insects especially, on a larger scale. These recovery initiatives may not come to fruition in the exact sense intended of the local habitat insects due to factors like the presence of alien species and the possible increase in nitrogen deposits in the area. The situation becomes even more complicated when discussing the recovery of these secondary forests because of the importance of epiphytes that include mosses orchids and bromeliads which are essentially organisms that grow on the surface of plant picking up moisture nutrients from the air and various debris that surrounds it because they serve as another linking integral factor in the whole ecological processes of insect habitats. We come into play when we discussing things like that would leave later that subsequently further down the line lead to healthier soil directly leading to healthier plants and again contributing to that integral whole that makes up the macroecological system.

So in a nutshell, when discussing the various components that function for tropical forests, insect conservation and protection the first look at intact primary Forest which is as far as we go into having the pristine natural environment as initially existed where all the native insect inhabitants would live and thrive and beyond that, we have the conservation corridor. The conservation corridor directly adjacent to it would consist of certain initiatives like barriers or even agro-ecology systems introduced so as to act as a railing to the intact primary forests. Next to that would be the regenerating forest where initiatives BIOS are consistently taken like growing the native fauna and allowing integration from the primary forest through the conservation corridor to come in and take hold of that ecological system. All these factors are important in the overall deliberate attempt to conserve, preserve and protect ; sometimes especially when looking at the regenerating forest even create insect habitats (Samways, 2019).

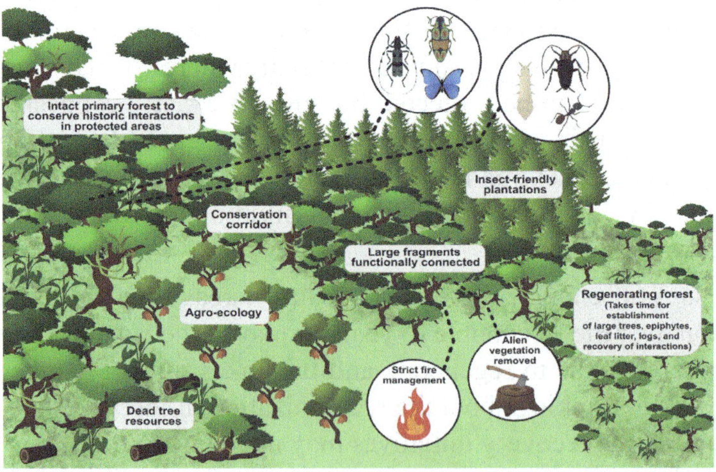

Figure 4.5: The paramount components for tropical forest insect conservation. Organizations and individuals must be patient as this is a slow and quite delicate process with the regeneration forest takes years and subsequently turns into the new conservation corridor as the preservation area is expanded.

Source: https://www.sciencedirect.com/science/article/pii/S0006320719317793

4.5 TEMPERATE AND OTHER NON-TROPICAL FORESTS INSECT HABITATS AND THEIR CONSERVATION AND PROTECTION

Temperate and other non-tropical forest habitats for insects may not be as complicated when it comes to all the factors involved however you will still find an intricate network especially when it comes to the structural diversity and it also still possesses quite a number of microhabitats that have their own internal microclimates delicately working together to build up specific delicate conditions for species to coexist. The North temperate region and the forests that exist there are not discussed here in relation to insects' habitat preservation and conservation especially, since they are highly reduced in size and some of them are largely modified with very little of the initial primary forests that used to exist there remaining in their natural condition (New, 2012).

The small-scale primary forests would have very old veteran trees and are a refuge sadly in some cases, the last for many native local insect species. There are especially some specialist insect species that live in a diverse range of microhabitat like those associated with deadwood which we mentioned earlier. Conservation initiatives go as far as the protection of individual trees with the philosophical mindset that they might just be carrying the valuable resource that we call the insect habitat. With insect species in Europe that are rare maybe endangered, you will find that abandoned farmlands have proven to be of incredible benefits because of the conservation interventions that have taken place with the Reformation of regenerating forests as a form of colonization credit. This situation again just like with tropical rainforest would require a long time: decades maybe even centuries especially, when considering the heterogeneity quality that the forest had some time back especially realizing the different insect species and the traits that they have.

What this means is that there may be intermediate stages in the ecological succession of the ecosystem when we try to regenerate the insect habitat back to its optimum levels and these are known as seral Communities or seral stages. What happens here is important and there needs to be a deliberate effort made into the functional connectivity through these stages allowing a chance for most if not all insect species to survive and even thrive. This requires an interesting and very delicate balance where conservation and protection efforts would create or foster natural forest gaps and various forest margins which in themselves are microhabitat but furthermore maintain the understory of the forest with a proper reference against the true history of the conditions that existed in the area and not simply along the lines of human beings and their intentions (New, 2012). It has been noted that this is one of the biggest challenges when it comes to the conservation efforts on temperate forests because they become so deeply modified that they actually depend heavily on human beings to manage them and without their efforts, the restoration processes would be futile. Some places for example in the US would have to set controlled fires to initiate or instigate the initial historical grassland or Forest design. Well beyond that there sometimes clever and well-thought-out does not rule 34 or large mammal cavities like those that bear a wild boar would create having to be mimicked. A technique that has been proved successful is the traditional grazing of livestock around these areas to create very natural forest openings while not forgetting the dropping the wood left behind acting as important elements for insects as well as the soil.

Log stacking procedures are also another temperate Forest management technique for conserving insect habitat especially, when looking at saproxylic species. Furthermore, pollarded snags, tree stumps and various forest debris like fallen branches are retained and introduced to the forest for insect colonization. Sometimes even wooden boxes are used because they mimic artificial tree characters that are again very beneficial to various rare saproxylic species. This is also delicately done with the introduction of indigenous trees and material to the local encouragement of insects and also to return the areas to its earliest known history or origin (New, 2012).

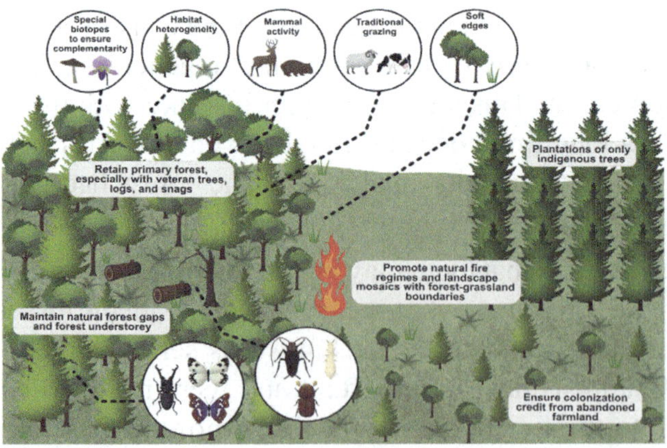

Figure 4.6: The chief components for temperate forest insect conservation. Note the involvement of natural gaps both to encourage the insects that use deadwood as habitats and also to create a diverse understory.

Source: https://www.sciencedirect.com/science/article/pii/S0006320719317793

4.6 LARGE GRASSLANDS AND OTHER LOW GROWING ECOSYSTEMS FOR INSECT HABITAT PROTECTION AND CONSERVATION

If anything the grasslands and low growth areas are the ecological systems across the world that have experienced the most impact from human activity in the course of History. This is especially evident in many parts of Europe with very little fragments of historical provenance still existing. Heathlands, sclerophyllous vegetation and various savannas have been cleared for human beings and their sedentary lifestyle but it has been realized over the last few years that it is important to leave some fragments and in many places maybe even tens of hectares have been left for insect habitats as truly they are the last remnants of rare insect species. Some places you will find that the tropical forests, non temperate or temperate forests will be right adjacent to these grasslands and they work hand in hand in providing the necessary resources which are associated with the insect requirements and needs and just like a forest region they will have to have and maintain a well-functioning connectivity (Samways, 2019).

The integration of all these ecological factors is something that will be similar in all biomes even up to the very smallest scales. Most of this is done even with grasslands by the creation of conservation corridors that will undergo several seral stages even if just one and this will help as a stepping stoneto improve the insect population and in turn, the habitat symbiotically. Which grasslands you will find that even well-managed roadsides can be very beneficial for various services up until power line services in urban areas because everywhere else has been either developed for human habitation or the exisiting grasslands are subjected to severe agricultural ventures. This extends even further true the rafts of golf courses, wind turbine sites, and airport areas where many grassland insects are able to find a habitat and a refuge.

We earlier mentioned the intensified agricultural initiatives that human beings have taken part in that I've taken away from the grassland habitations that insects have however a moderate level of livestock grazing especially when you have well-fine-tuned techniques like rotational grazing regimes you will find that you are able to create a habitat heterogeneity for local grassland insect species. Furthermore, they are proper machinery and mowing techniques that can be used which will not damage insect grassland habitats beyond their need (Samways, 2019).

It has been studied well indigenous grazers even livestock are present there are more effective heterogeneity and functional connectivity created for insects' diversity to be increased and this has been observed in Africa. In conjunction with that controlled fires and natural fires in the natural periods of intervals also improve the habitat but if this is done on an artificial level then there must be appropriate integration so that opportunities are created more evenly for greater insect diversity. This can even involve bringing in or reintroducing megaherbivores that used to exist in these areas before the degradation took place. It is especially more technical where hydromorphic grasslands exist with periodic flooding being a necessary occurring event at its intervals and may be artificially done to improve habitat quality. This is done in many areas of South America where there are extensive Wetlands in the Pantanal. Sometimes even careful delicate disturbances of the soil surface such as light plowing can be used to improve the various levels of sward. Various grasslands are sensitive especially to any additions of nitrogen even atmospherically and more so when it is introduced artificially with the excessive use of fertilizers. Nitrogen candy removed when it is an excessive quantity from Wildlife and insects' reserves by intensive mowing or grazing although both keys activities as we said when they are intensive will be detrimental to the local insect species. Pesticides also must be avoided and these areas even wear the appear not to cause a major threat. Some of them can be harmful to domestic livestock like Ivermectin but can also affect dung beetles in lethal fashion.

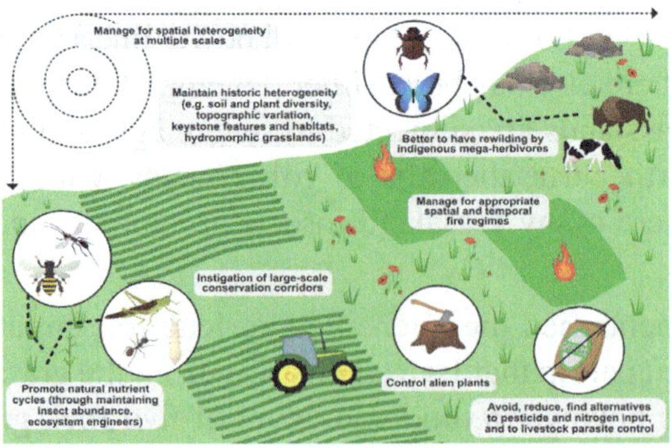

Figure 4.7: The core components for grassland insect conservation. Note the involvement of wide corridors especially where aggressive agriculture is done. Also, observe the controlled fires that are involved to reset the local ecology.

Source: https://www.sciencedirect.com/science/article/pii/S0006320719317793

4.7 FRESHWATER INSECT HABITAT AND THEIR CONSERVATION AND PROTECTION

Freshwater makes up less than 1% of the earth's surface but serve as a habitat or simply support in one way or another. 6% of all known insect species and they are an integral part of the general topography biogeochemistry, hydro period and general vegetation in the water and along the banks of these freshwater systems. Wetlands have been reduced over time and for insects, the major concern is pollution and agrochemicals which go further in affecting biodiversity in general. Many protected areas have been built up and they play a big part in that you will find that aquatic insects and the preservation are not within the major concerns- super major aquatic insects are left out. With terrestrial insects, there are extensive pieces of land especially for us that are still intact that in turn could be essential for many aquatic insects. You will find such plants like bromeliads and pitcher plants being very important for aquatic insect species and even support rare species of mosquitoes are swell but overall biosphere reserves have actually played a major role in conserving aquatic insects (Samways, 2019).

The wide catchment areas and wetland conservation or freshwater systems have played an important role in meeting the aquatic insect management objective. Within our local environment or at the local level streams and their subsequent conservation and preservation has been particularly important especially when considering endemic insects that usually exist in high-elevation ponds. These areas need to be maintained especially, towards there provenance settings where the dynamics will be historically at par with what they used to be including aspects like seasonality and the basic physical chemistry of the water conditions. Another important factor is producing other location- specific factors that would be in part sensitive to the topographic conditions and even various biotic interactions like promoting the macrophytes and the bank vegetation quality that exists with its local diversity.

Channelization which is when engineering Tactics are used to straighten or redirect natural streams, artificially modifying them should be avoided because it greatly reduces the insect diversity in the area and in some instances can actually be the underlying Factor in local hydrological drought. On the other hand, you will find that artificial shallow shorelines especially those that are diversity vegetated can be used to create as well as maintain underlying layers especially of rock or soil beneath the surface of the ground in wetlands that can be rich in organic matter and subsequently raise the insect diversity in the area. It should also be noted that local integration of fauna and flora like certain vertebrate engineers for example, beavers should be recovered or at least reintroduced into the area. Sometimes, if it is not possible then alien trees can be a decent substitute for the loss of the historical local trees but it should be seen as a temporary measure because studies have shown when they are removed it leads to noticeable insect habitat improvement. Floodplain areas should be maintained as well and the gravel areas around them should be left intact because they are important for both terrestrial and aquatic insects. Saline levels should also be a consideration when it comes to the conservation of wetland insects' habitats (Didham et al, 2020).

As we have seen with the other congregation strategies in various other biomes connectivity must be involved with all parts interacting seamlessly. With Wetlands or ponds clips the idea should be to improve heterogeneity and size variations not only in the individual Wetlands in the area but with all neighboring ones the streams and rivers as well. That must be the natural historical dynamics continue to play out as for both vegetation and insect diversity. It should be noted however that they are sound insect species that are adapted for certain conditions like short hydroperiods and so they may be a necessity to have a variety of both permanent and ephemeral ponds so that streams are able to deposit all sorts of materials again adding onto the heterogeneity in the area. Some aquatic insects are more designed for permanent wetlands and pools or forms and so to colonize these areas. There must be an instigation of the environment and maybe the introduction of permanent artificial pools are necessary. They may be able to provide these conditions by using high-quality irrigation ditches more so for Marshall and insects and even stormwater ponds for those insects that are more aquatic as long as the physiochemistry of the water is unaffected.

Natural vegetation adds to the connectivity in wetland insect habitat conservation in both Agricultural and urban ponds and insect phones have the added benefit of increasing awareness about aquatic insect life the habitat and the conservation. You will find that some human or artificial design landscapes can actually provide a great opportunity and a great habitat for aquatic insects the preservation and conservation and they include variations from garden forms to military training areas and even roughs on golf courses where water exist (Didham, et al, 2020).

Some more unique or special cases that are notable in terms of aquatic insect conservation include reducing shipping or both wave impact and sometimes the introduction of biological control of invasive water plants. Because terrestrial and aquatic areas are very distinctly different: they fall in what we call in the ecological word and ecotone which is a transition area lying between two biological communities or environments in which they both meet and integrate. They could be very local or regional they could be very narrow or wide but in this specific case they are more or less quite local and these areas especially when looking at river-lake Echo tones or gravel areas from wetlands into woody grasslands: it involves the conscious and deliberate effort of conserving them in the way they behave and exist. This could involve going so far as to remove alien predators from the environment such as fish that did not exist that would feed on the local insects. The erection of physical structures has to be well thought out as well as they may interfere of threaten the flying adults' period: in some areas, tourism actually has a negative impact and especially around more special and sensitive habitats and has to be well thought out as well.

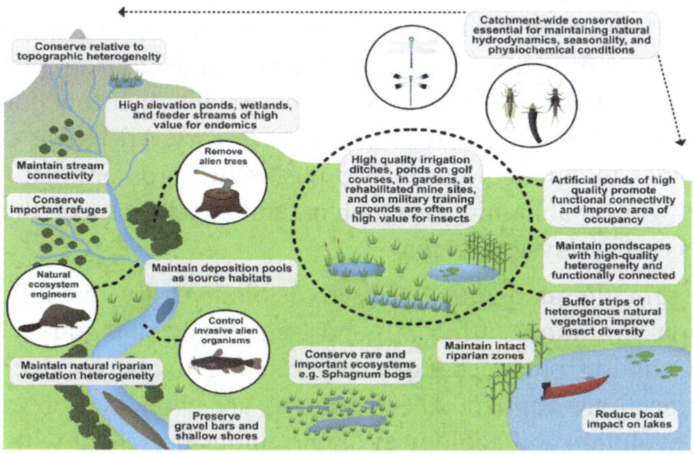

Figure 4.8: Essential components for freshwater insect conservation. Note the involvement of local flora and fauna as the alien species or both are worked against.

Source: https://www.sciencedirect.com/science/article/pii/S0006320719317793

4.8 INSECT HABITAT CONSERVATION AND PRESERVATION IN AGRICULTURAL LANDSCAPES AND ECOLOGY

Agro-ecological Tactics and strategies for improving insect habitats for the preservation and Conservation involve a paradigm shift philosophically and practically. It involves land sharing and land sparing dichotomy where both productions of agricultural goods and conservation of local animals and in this case more especially insects are well-integrated. This involves a total mixing of production plants and natural vegetation which is what is known as sharing and also through actual special separation from the production areas to the designated conservation areas which is known as sparing. The land sharing end involves bringing in historically indigenous trees that would for example, provide shade for the production plants like cacao or coffee while consequently providing a variety of insects with both food source and habitat. You will also find that naturally occurring local flowering plants will provide local pollinators with the resources that they require and they ,in turn, will control agents of aphids and other undesirable flora. When looking at land ,there should be large scale networks known as corridors or conservation corridors that are set aside from plantations to allow a wider range of insect species to coexist with other functional groups. These areas could be fenced out as well if necessary. All Things considered, when it comes to agroecology is also pushed towards optimal production on one end while maintaining as much natural biodiversity on the other end creating its own ecosystem in that.

High agricultural yield , effective biodiversity preservation and conservation are more effective and seamlessly integrated when the traditional cultural practices especially relating to the local crop that is being produced are applied. These systems are employed correctly could have both evolution and ecological opportunities presented with long-term sustainability guaranteed because you have safeguarded to some capacity, the ecosystem itself and its self-renewal mechanisms (Kumar, Yogi, &Jagdish, 2013).

Other factors to consider when looking at agroecology is the insect species richness abundance and the coexisting assemblage composition. Crops will require fluctuating pollinator abundance that effectively drives the local ecosystem with an abundance of certain insect species to be more important than others. For pest control considerations, you will find it will be more effective to have a few species of predators of parasitoids that will control the pest species. This becomes more intricate when dealing with spatial separation of conservation areas and generally speaking the idea here is to provide opportunities for both the common insects and the rare species across the local scale right next to the commercial crop and also in a wider landscape the area.

Agro-environment schemes which were first developed in Europe and later on spider cross the world are set aside land production areas which are beneficial to the local diversity of fauna and Flora and would involve maintaining field borders and hedgerows so that a higher number of species in their diversity is encouraged especially outside the production areas. These agro-environment schemes can be beneficial to flower- visiting insects for example that feed on their pollen. Because these areas would be pesticide-free they are a great breeding ground and so would require some sort of management where herbivorous and parasitic insects would be controlled and this is mainly done by the improvement of field borders and the establishment of conservation Headland which are basically margins that would reduce the amount of pesticides that is to be used. Plant-rich roadsides and buffer areas that sometimes could even involve ponds can also improve habitats for insects especially, where agrochemicals are being used which could have effects on them (Kumar, Yogi, &Jagdish, 2013).

Cereal fields and their conservation effort in headlands would have bumblebees that what do you use the pollen and nectar resources and also help the plants flower. If one has good knowledge of the ecological requirements of bees ,in general, there could be an overall result of new species having a great life here. Also, when you have hay Meadows that are left to their own devices also increase insect diversity and field margins that are minimally maintained and have no inorganic fertilizers will improve local insect lives.

Organic farming is another way that agroecological systems can be developed with its main aim being to move away from negentropic and artificial substances that would affect any functions of conventional agriculture. In essence, this works by lining everything with nature for a healthier and more resilient landscape. Organic farming is beneficial to insect diversity when compared to general large-scale conventional farming with a 30% increase in species richness and comparison and also serves the local historical species of insects. Organic farming can prove their effectiveness by also involving land sparring techniques but this may reduce productivity especially when looking at highly productive cereal fields. When groups of organic farms are organized together to increased functionality and the well-connected and operated in conjunction they will have both higher yield and ,in turn, will increase the insect diversity in the area. Organic farming can also be beneficial in the mail hold horse habitat for local insects increasing the diversity across the conventionally managed fields that may exist near them providing a countryside wide opportunity for richness and abundance.

The heterogeneity of the landscape will also improve insect pollinators and the diversity of them sometimes even generating wild bee species and this can symbiotically work in producing high but appropriate floral diversity as well which means monocultures with their adverse effects are mitigated. This brings up a whole philosophical discussion as we have been propagating landscape diversification which further becomes a factor especially, when combating overall climate change. All this must be thought hand-in-hand with the natural habitats that do exist producing a mix of nectar and pollen seeds ,for example, to encourage their rabies to restore the historical environment.

Crop field size is another factor when considering how to widen the heterogeneity in the environment because crop field edges bring a more complex fauna and flora interaction in the area and this may be why smaller crop fields are better considerations especially, when practicing organic farming even though the yield might be compromised when natural and local crops are being used and remnant patches are left to their own. Furthermore, woody habitats that lie adjacent to grass and patterns in these organic Farmland areas further improve the insect richness abundance and diversity by providing an assortment of complementary resources.

4.9 INTEGRATED PEST MANAGEMENT AND THEIR FUNCTION TOWARDS INSECT HABITAT CONSERVATION AND PRESERVATION

One thing that should be made clear is that many pesticides pose a risk to human health but furthermore they distract all sorts of biodiversity that exists and in the long run you might find that some of the pests that are even trying to be eliminated become genetically resistant making the pesticides ineffective and in turn forcing human beings to produce even more lethal and toxic pesticides which in the end becomes a vicious cycle. Neonicotinoid is an example of an insecticide that actually severely affects bee behavior which could end up rendering the flowering and pollination processes that would otherwise naturally occur.

This means that overall they must be some sort of Integrated Pest management techniques and methods that should be used which traditionally even used to occur. Methods such as intercropping were used historically as biological control measures against pests which had a further effect in improving the insect diversity in the area. Using trees to provide shade will also increase the activity of predatory insects in the area especially where cacao is involved with cacao pests and predaceous ants. There are other techniques which involved reducing field size and increasing the density of glass for example which would encourage the parasitic wasps *Aphidius colemani* that are probably the most destructive insect pests on cultivated plants especially, in temperate regions by detecting the aphids honeydew secretions while also sensing distress signals from infested plants.

4.10 CREATING CONSERVING AND PROTECTING INSECT HABITATS IN URBAN AND SUBURBAN AREAS

Of course, cities and towns have high human density and well- modified ecosystem that may not be well inclusive of insects but many through the creation of green spaces can host great insect population and play their part in insect habitat preservation and conservation. As much as green spaces may not be a substitute for wild areas some of them can be very rich in a variety of insect species especially in semi-rural areas in many parts of the Western world. This input would help especially where herbaceous and rubbery vegetation exists with more green spaces than hard spaces and having a good set of large trees. The idea here is for more successful development plans must have larger purchase of land with as much structural variation and diversity in the flora that exists with places where especially aesthetics is not important being left to function naturally as it would. Public parks ,for example, could fit in here and sometimes may act as great conservation and protection areas for rare species when they are rich in the resources that they need.

They are of course quite a few challenges in this respect like artificial lighting ,for example, but there are solutions to this. Metal halide lights are a bit detrimental insects and they can be replaced with Sodium lights and light -emitting diodes. This can be further enhanced by having low-to-the-ground lighting along paths rather than structures that illuminate wide areas unnecessarily. This could of course involve having some energy -saving techniques combined into this strategy like having sensor lights or turning off the lights during daytime and having them on late at night and maybe partially in the early mornings.

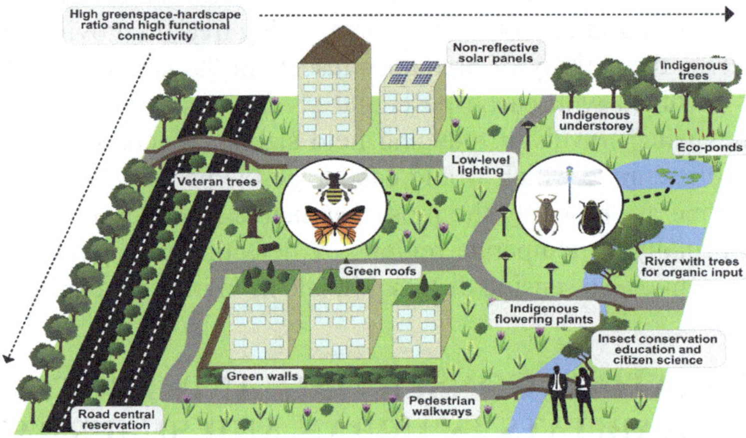

Figure 4.9: The essential components for urban insect conservation. Observe how veteran indigenous trees offer a form of habitat for a set of species while eco-ponds may offer another and local flowering plants having yet another set of insect habitats. Green spaces are an underrated yet very versatile form of insect habitat conservation and preservation.

Source: https://www.sciencedirect.com/science/article/pii/S0006320719317793

In conclusion, with what has been often referred to as the great insect dying they have been emerging themes for insect recovery over the years and although there are challenges posed it must be recognized that biodiversity loss climate change and human well-being are interconnected as a whole and there must be strategies developed to address all these issues from an interactive an integrated perspective. Insect conservation progress he is in itself our whole philosophy of valuing nature and back stent respecting ourselves in the end and the historical -cultural heritage.

Although it is evident that we are not going to be able to preserve every insect population and even every species due to a naturally occurring Extinction that is bound to happen the decline of insects and its deep Impact for planetary survival is being recognized by the general public and it has been evident with the concern over the loss of bees and pollination agents across the world having remedial initiatives taken in that respect. Insects are integrated into our ecosystem and ecology and their predicament is bound to be integrated as well to larger global effects like climate change and loss of various other biodiversities. This means that each of us has a role to play in insect habitat preservation conservation protection and creation.

5

Implementing Suitable Systems for Monitoring Insect Populations

CONTENTS

5.1 INTRODUCTION

Estimates from the conservation of insects make the suggestion that about 50-90 percent of the insect population is yet to be discovered but still those that are known make up greater than half of all the known animal species. Currently, there is a poor knowledge base on insects given that most species are yet to be discovered, monitoring changes in the diversity of insects poses a challenge to scientist and their attempt in the generalization which involves large extrapolations from the few species which have been studied. One of the groups of insects that is mostly used to measure accurate changes is butterflies. Ecological monitoring is useful in providing the estimation of the abundance of insect species and this is important in making decisions. It is ,therefore, necessary to understand insect `ecology before making decisions related to monitoring insects.

Understanding insect ecology can be helpful in monitoring the population of insects. Insect ecology is the scientific study of insects as an individual or a group interacts with the environment with their environs or in their ecosystem. Insects play an important role in the ecology given their considerable biomass and their interactions with plant life, other organisms plus the environment. In the case, it can be defined as the ratio of insects that are beneficial in relation to those that are considered as pests and their distribution patterns over time. This knowledge is important to assist farmers in making decisions in relation to pest management. Insects are valuable in both terrestrial and aquatic ecosystems and are an important link in many food chains and webs.

Insects form a valuable part of the biodiversity. Changes in the population of insects can be detected in other levels of the food chain and monitoring their populations allows scientists to predict effects on animals at higher levels in the food chain. Insects have been used in the documentation of long term changes in habitats and the record of their fossils has been useful in the construction of climate history. Insects have a short generation time and together with their mobility, they are able to respond rapidly to any change in the environment such as a change in the climate, fire or the exotic outbreak of species. Therefore various communities of insects or other invertebrates can be useful in the detection of environmental changes such as pollution and changes in the climate which means they can be used as ecological indicators and can offer better understanding in relation to the changes which occur within an ecosystem.

Insects respond to climatic changes because of their specialized habitat requirements. Butterflies are insect examples that have shown responses that are rapid to changes in the climate and the implications can be serious ,for example, the montane butterfly where a significant reduction has been predicted especially when there are drought conditions.

Fire is a factor that affects insect populations especially above -ground insects of which effects that are direct include death by incineration in comparison to those insects found underground such as beetles and other burrowing insects that are less likely to be affected. Indirect effects of fire can be observed on if vegetation habitats for insects become affect by fire. For insects, whose survival depends on dead wood, an increase in their populations is usually observed for example pine bark beetles.

Introduction of exotic species for example plants may have large effects on terrestrial insects and sometimes this may remain undetected. Exotic species are sometimes transported by human beings and some of them include the knapweed, Canadian thistle and the leafy spurge and they can be useful in an indication of the health of an ecosystem. Their effects can be indirect and can affect the number of insects and it can also affect plant and soil nutrients that are available to insects. Insect populations may experience changes for example extra nectar, food and host plants, there can also be a reduction due to preferences in relation to nectar, food or the hosting plants.

The list of species that are to be monitored should depend on the following the potential of the insect group in relation to the functioning of a given ecosystem, knowledge of the species or the group, potential of the species as an indicator of biodiversity and finally the availability of standard and cost-effective methods of monitoring insect populations. Insects have an influence on the productivity of plants and the cycling of nutrients. The outbreak of invertebrates can make contributions to change in ecosystems for example the vegetative matter especially in the climate that is changing.

Monitoring insect populations in terms of their abundance is also important in understanding their dynamics, planning and also the assessment of insects that are targets of conservation. Most of the time monitoring is defined by space and time. Insect monitoring involves using traps in order to gain information on the activities of insects. If we know the life stages within the functional groups, we can see the past trends which may also continue into the future (Barbour, 2016).

Insect monitoring is a great way to save money and reduce the use of pesticides. There are different insect functional groups in different ecosystems. They include pests, generalist predators, key predators parasites, key parasites and decomposers. Their populations tend to increase or decrease depending on different factors such as climate, food, supply, predation and many other factors. The reproduction of these functional groups does not occur at a constant linear rate.

People monitoring insects for the purpose of pest management, count the number of key pests individually in order to predict future infestation. This form assumes that insect species continue reproducing in similar numbers as in the past but contrary to this they produce exponentially if they do not experience predation. Managers are required to make observations on the functional groups in relation to each other to gather patterns and make decisions. To make proper decisions on the management of pests we need to observe the demographics of different insect functional groups.

Pests, even a small portion can be an asset to all farms. Pests can be a source of food to both primary and secondary predators and are therefore important for the reproduction of many primary parasites. Secondary food sources for beneficial insects or other inset s other than pests may also be of great benefit to the grower. Different farms will have different pests, alternative sources of food and beneficial insects. This means, there are no standard values that are used to determine the actual ideal numbers that can be required in sustaining an effective natural enemy complex on a farm. Throughout time, the pest and other functional group relationships should be observed so that there is the determination of control and reproduction naturally (Barbour, 2016).

The ecology of insects can be observed in ways by sampling and making observations the greatest number of insects in an area. Monitoring the number of key pests does not give sufficient information that can be used in decision making. Samples should be taken from areas where insects are abundant or where they appear first. The objective when sampling is to make observations on as many functional groups as possible so a ratio that represents the different functional groups can be observed.

Monitoring can be done during the growing season especially when there is a sign for pest outbreak. Some of them will be favored by patterns of climate that are warm such as counting degree days and observing codling moth. Some will also successfully invade farms during certain stages of plant growth. Microclimates which provide humidity may be helpful in the determination of the presence of functional groups. Insects such as ants create a space that is free of enemies by searching for honeydew on plants and a sign that those involved should continue monitoring their activity closely. Recognizing pests and beneficial insects is the basis for monitoring insects in a manner that is efficient and it is ,therefore, the first step in pest management to avoid unnecessary use of pesticides. There are different tools for monitoring insects. Most of the time traps are used to provide a value of the population density in various locations and this also comes with its challenged and will be discussed further (Barbour, 2016).

5.2 SPECIES AND POPULATION MONITORING

Species and population monitoring are observations made on a regular basis and records made on changes in status and trends in species or their abundance in a particular area. The primary objective of monitoring is to get information that can be used to analyze the outcome of management actions and to also guide decisions on management. Assessment is done on species that are in danger of extinction or when there is the need to make the existing ones more intense. In the case of pests, it is done to assess the management action that is to be taken. Variables that can be measured in the monitoring process include changes in the abundance of species, trends in population size and structure, changes in genetic diversity, predator numbers, spread of invasive species, vegetation cover change and effects of management interventions.

5.2.1 Trends of Insect Populations

A decline in insect populations has been observed globally. They are a variety of factors contributing to this to this decline. This depression affects a variety of insect species the pollinator insect's species, pest and parasites alike. Some of the most affected species include bees, butterflies, moths, beetles, dragonflies and damselflies. A study done in 2014 by Rodolfo Dirzo, showed that they were a decline of 45 % in the population of invertebrates over the past four decades and out of the 3623 species of terrestrial invertebrates as identified by the international union for conservation, 42 % of the species were threatened with extinction. This decline is far from uniform as in some areas that there is an increase in the overall population. On an average, the decline is said to be at 9% for every decade those this figure may not be a global representation.

A survey done on threats to insect life in London concluded that there is a severe decline of insect populations and this also affects other animal populations that are higher in the food chain and also affects the supply of ecosystem services like pollination and nutrient cycling. In countries like the United States and those in Europe have recorded a 40% decline in the population of bees due to colony collapse disorder. Insect species like the monarch butterfly are also experiencing a decline in their population. Scientists have observed that this decline is happening gradually and many people may notice the effects of this when it is already too late (Medeiros et al, 2013).

5.2.2 Factors that Limit the Abundance of Insects

They have cited different factors for this gradual fall in the population of insects but the greatest one in the continuous use of pesticides, the spread of monoculture crops, urbanization, industrialization and the destruction of habitats. The different species of insects may not be affected in a similar manner. Most groups of insects are subject to limited research and lack of figures for comparison from earlier decades may also not be available. Errors in the analysis of data and exaggeration of the decline of insect populations are attributed to a decline in the field of entomology. Conservation efforts have increased in response to the decline of insect populations, for example, the German government in 2018 developed a conservation program called Action Program for Insect Protection and a group of entomologists and ecologists have called for action to establish a research Centre to allow for research on insects and how their decline affects ecologies.

Competition: Competition for available resources is also observed to increase with increasing population size. Competition among species may lead to the elimination of some species. As competition increases, there will be an increase in the species that are dominant and in some cases, there is continuous coexistence of the species for example density -dependent and nonlinear density- dependent factors which operate respectively. Parasitoids, parasites and pathogens are factors that are dependent on density factors. The effect will vary depending on the physical quality, time factor and spatial characteristics of the biotype and finally changes in the composition of species within an ecosystem.

Parasitoids and predators can lead to a restriction on the growth of insect populations. They can sometimes act in a manner that is different depending on the host population. Insect conservation biology suggests that parasitoids and predators are also subjects themselves and they are impacted on with the loss of the host, mutualists as well as parasites and predators. Another consideration is predation by vertebrates is also an important consideration. Introduction of insect predators and parasites are regular subjects of successful classical biological control while the introduction of vertebrate predators can lead to catastrophic impacts upon insect populations (Medeiros et al, 2013).

Insects are vectors of plant diseases. Most causal pathogens are either viruses, fungi or bacteria. These diseases affect the competitive advantage of some plants and this has an indirect effect on the conservation of insects by altering microhabitats and even biotypes. Insects can also be subject to pathogens including viruses, rickettsiae, bacteria, protozoa, fungi and nematodes. The degree and the type of infection will be determined by the type of weather. Mortality will vary from one population, a single species to another. Factors that limit the abundance of insects are complex and vary depending on the species and the interactions within an ecosystem.

Anthropogenic factors: Ecosystems are under pressure as a result of human activities and this has increased over the past years. The pressure created goes on to affect the structure of ecological communities and most of the time it also stimulates the outbreak of destructive species and this leads to the collapse of top-down control. Biological invasions are one of the major reasons for the degradation of ecosystems and the loss of the world's biodiversity. This situation is made worse by changes in climate and also the environment and this has the potential to affect the abundance of insects.

Abiotic factors such as weather and climate reduce the insect population regardless of the population density. The population dynamics of insects is directly related to climatic and weather conditions which go on to affect the physiological processes of a living organism. Weather leads to the variation of insect populations over time and this varying change in insect abundance can affect the efficiency of the different methods used in monitoring insects.

Changes in ultraviolet radiation as a result of ozone layer destruction can also affect the response of insects. It is easy to destroy the viruses of insects and therefore will lead to the inactivation of the rate of viruses found within the environment and can be important in influencing the population dynamics of insects (Barbour, 2016).

Comprehensive monitoring of the ecology is necessary in order to provide information that is sufficient and timely especially those that are going extinct and those that have the potential to cause problems ,for example, monitoring new species which may become harmful pests a good example of the latter is the gypsy moth in the united states and the Japanese knotweed in the United Kingdom and the *mnemiopsisleidyi* in the black and Caspian seas.

In terrestrial ecosystems for example places where agriculture is practiced, ecological monitoring is usually part of the integrated pest management program which advises the use of pesticides in the event that exceeds a certain threshold. Using pesticides has its disadvantages such as additional costs such as operating costs, additional damage to crops and potential environmental damage.

Consequences of insect population decline could have far- reaching consequences for nature and also to human beings who depend on insects and other invertebrates for the pollination of crops ,for example, a study that was published in Canada in 2010, suggests that the North American bird species depend on insects in the air for food and that their young ones suffered much more visible declines in recent years, unlike other perching birds that commonly feed on seeds. This analysis is based on data from the North American breeding bird survey (Barbour, 2016).

The root causes of insect decline are yet to be fully understood but most scientists agree that they result from a combination of factors like the destruction of habitats, deforestation, fragmentation, urbanization and agricultural conversion. Immediate solutions to this problem include reducing the use of pesticides, cultivation of a variety of crops and reducing water and light pollution. In addition to these solutions, the road map suggests giving priority to species that require protection by conservation.

To better understand this problem, scientists is urging an increase in efforts to monitor insect populations. A recent increase in monitoring efforts has risen from citizen science projects where individuals who are non-scientists are trained on the collection of data an example is the butterfly monitoring project which runs in association with the butterfly conservation project in Europe. They collect data on butterfly populations annually. Globally, they are a lack of comprehensive data on the various comparisons and therefore the existing data is limited and is restricted only to a few groups and only a few locations.

This creates the need for increasing monitoring efforts at a level that is large and also on a global scale. Efforts are also being made to develop automated monitoring systems in addition to the conventional methods. Many scientists are in support of the action to increase the monitoring of insects and other invertebrates. They point out that there is already enough existing knowledge on insect decline to start addressing the different causes manly in agricultural policy ,for example, new attempts are necessary in order to increase incentives for enriching landscapes with hedgerows, reducing fertilizer and the use of pesticides and better rewarding organic agriculture because previous efforts to achieve this have failed in the past. According to Deckert, one of the key questions is the view of governments on biodiversity, whether they see it as an odd or its existential importance for future populations (Barbour, 2016).

5.3 INSECTS AS BIO INDICATORS IN MONITORING

Biomonitoring or bio indication is a type of applied ecology. Its primary goal is to use organisms living within natural communities to monitor the impact of disturbance and use this information in managing ecological ecosystems. It can be applied in both urban and agricultural communities. Biomonitoring has existed for a long time since 1908 when two workers in Germany published a paper with the use of macroinvetebrates in the assessment of the quality of water and the level of pollution in rivers that were subject to contamination by sewage waste. The use of biomonitoring did not start until the 1970s. Factors that limit the abundance of insects are complex and vary depending on the species and the interactions within an ecosystem.

5.3.1 Levels of Bio Monitoring

There different categories of indicators. They include environmental, ecological and biodiversity. Each of the categories has their function in monitoring.

- Environmental indicators- they are used to indicate changes in the state of the environment. They are used to monitor changes in the environment. The species used as indicators in biomonitoring respond in a predictable manner and in a way that makes it easy to make observations and quantify disturbance in the environment. The organisms are used as devices for providing an early warning or to reduce the effects of a disturbance for example an effluent. They act as surrogates for the larger community. Organisms can be used as accumulators of chemicals that are used to bioassay the presence or the concentration of pollutants. The organisms are used to rank pollutants in order of the level of toxicity.

- Ecological indicators- these indicators show the impact of the stressor on the biotic factors and they monitor changes that are long term in relation to the stressor. Rather than merely functioning as gauges of changes in the environment and they are used in the demonstration of the effects of altering a habitat or making them fragmented or climate change among others. Their response may be a larger decrease in the population size, the spatial distribution or changes in life history. The response observed is similar to that of other species that fall within the same taxonomic groups that exist within the habitat.
- Biodiversity indicators-they are used in the identification of diversity of taxonomic groups in a specific area. They are used to monitor changes in biodiversity. Biodiversity indicators a taxa group whose diversity is representative of some measure of the diversity of other taxonomic groups in a particular area. The diversity measures could include the richness in characteristics, species, level of endemism, genetic diversity among other measures. The diversity measure of the taxonomic group used as an indicator can be used in the measurement of the diversity of other species.

Bio monitoring can be done at different scales at an individual level, among populations, community level and ecosystem level. Direct measurements can be made of physio-chemical factors of an environment. In terrestrial ecosystems, it is common to know the type and the degree of changes in the environment for example fragmentation, the introduction of a new species among others. Most of the time it is difficult to monitor every aspect of an ecosystem. Another problem comes about as to whether the scientific community is aware of all the important aspects to monitor. In ecosystems that are aquatic, the problems encountered in monitoring can be much more complex in comparison to terrestrial ecosystems ,for example, assessing water quality requires knowledge on the hundreds of substances that can be monitored.it also requires understanding the synergistic effects when two or more pollutants interact and finally continuously monitoring aquatic environments for the detection of pollutants whose release is periodic such as the release of industrial waste at night. We know what we are measuring before it is measured. These types of monitoring usually consume a lot of time, labor and other resources. Some of the synergistic effects sometimes can go without being measured. The stems around the fact that the knowledge we have on natural ecosystems may not be sufficient and this makes us unable to make predictions of how disturbances will have an impact on various ecosystems (Campbell, Mullen, and Dowdy, 2002).

There are several advantages to using insects for biomonitoring. Many taxonomic groups are different with regards to how sensitive they are to changes in the environment and their habitat requirements and therefore the groups of insects will be selected according to the solution that is required. It will be easy to focus on functional groups such as primary consumers or top chain predators to monitor the functioning of ecosystems. There is also a lack of ethical constraints in the collection of insect samples for monitoring, there are no laws on the damages caused to insects in the process of monitoring. Insect populations are usually very large and the death of a few species does to have an impact that is negative on the insect populations. Insects can function as canaries for damage to the environment that can cause harm to human beings such as the quality of water or toxic chemical build up. The primary goal for monitoring the environment is to ascertain the effects of the disturbance on life and the use of species that are alive in a direct way.

Several factors determine the distribution of aquatic insects. They include the presence of oxygen, temperature, the type and nature of sediments together with substrates and finally the presence of pollutants such as pesticides, acidic materials and heavy metals. Following a disturbance, aquatic insects will vary in their response. The increase of materials such as sediments can increase the population of certain insects such as mayflies which have abdominal gills that are protected and other insects such as caddisflies for example those that are filter feeders increase in relative abundance. A reduction of dissolved oxygen, hemoglobin possessing blood worms have an increase in their population. As temperature increases the nymphs of stoneflies decrease. The diversity of species reduces when there are pesticide runoffs. Water that contains fertilizer runoff or livestock waste can have an increase in the level of nutrients will lead to an increase of some species and a decline in diversity will still be experienced (Barbour, 2016).

In terrestrial ecosystems, insects can be used as environmental indicators. In the past soil, insects were used in indicating soil fertility and pollution levels. Nowadays it can be used in a variety of situations such as measuring changes in the environment as a result of interactions between biotic and biotic components. The biotic community can have continuous change even after traces of the impact are no longer measurable. Studies that make use of biological indicators mainly focus on species that are in danger of extinction most of the studies have been done are related to butterflies.

Biodiversity indicators are used in the estimation of diversity. It is a faster and less expensive way than carrying out comprehensive biodiversity surveys. Only a few taxonomic groups of insect have been identified to date and they be efficient biodiversity indicators as science rapidly expands ,for example, tiger beetles are a species that have the potential of been good indicators or regional biodiversity patterns because they have a taxonomy that is stable, they have biology that is well understood, life history, the ease in taking samples, their global distribution and finally the specialization of individual species within habitats. Other potential taxonomic groups that can be used as indicators include scarabaeid beetles. According to brown insect groups with desirable qualities include Collembola, Formicidae, butterflies, isopteran and Coleoptera. The use of a single species to measure biodiversity has often been challenged and they are sometimes referred to as predictor insects. The predictor sets would ideally encompass different ecological functional groups for example a pick can be butterflies, termites, ants and dung beetles in the estimation of biodiversity in a forest for example. In this case, butterflies are phytophagous. Termites and ants are the most abundant invertebrates in the soil and canopy respectively. Termites and dung beetles are decomposers. The disadvantages of using a set of predictors are the number of resources consumed in the process.

5.4 INSECT MONITORING AND INTEGRATED PEST MANAGEMENT

Attributes that are monitored include quantity, population dynamics and the population structure and finally the habitat requirements. Demographic monitoring is a form of monitoring populations that assesses changes in the population such as mortality rates, growth, size, density and distribution. The information can be used to determine the distribution and species abundance in the prediction of the future abundance of populations.

Genetic monitoring is used in the assessment of genetic diversity for pollutions that are targeted in terms of their richness or their evenness. This information can be used in the comparison of populations in order to determine which species and determine the population that will be selected for conservation. It can also be used to determine the genetic flow between populations (Barbour, 2016).

Habitat monitoring is used to assess changes in conditions of an ecosystem or habitats that have been targeted. This covers a range of aspects that can be monitored such as the distribution of the types of habitats, an abundance of species in the habitats, composition of species in the habitat. Features include quantity in times of area; quality in terms of composition, structure, dynamics and functions in relation to ecosystem processes.

5.4.1 Insect Monitoring for Integrated Pest Management

There is no possibility of completely destroying pest and therefore monitoring and the management of the level of pests is advised instead of getting rid of them and this can be instrumental in environmental preservation, reduction of costs and protection of health for both humans and animals and still be helpful in the maintenance of organisms that are beneficial such as bees, bugs, birds, butterflies and pollinators.

Integrated pest management is an insect management approach that uses the knowledge base on pests, their life cycles, cultural practices, non-chemical methods and pesticide use in the management of problems related to pests. Including the contributions of insects that are beneficial to pest management is not easy. Identifying the insects that impact on pests that attack crops and estimation of their impacts they may have is important in determining whether beneficial insects are less likely to have any impact on the population of pests below the threshold, or whether a person will need to treat attack by crops to prevent the loss of crops (Barbour, 2016).

Integrated pest management has three essential program essentials they include, identification, monitoring, record-keeping, action levels, criteria of tactics and evaluation of the program essentials. Prevention measures are added to the existing structure as a method of primary prevention. Identification of pests because they will exist in different ecologies. Monitoring entails regular inspection and trapping to determine the type of infestation and their populations. Record keeping is done to establish trends and patterns in the outbreak of pests. An action level is the population size which requires remedial actions. Tactics criteria mean that chemicals should be used as the final solution after mechanical, cultural and biological approaches. The last step is evaluation and is an essential step to determine the success of the management process applied.

The monitoring process is designed to provide accurate and timely information on pest activity. It is done to determine whether there is a pest problem in the first place and also identify the causes. A schedule can be put in place and a plan for monitoring pest populations in order to ensure successful and effective pest control. This is helpful in determining acceptable the population levels of pests, effecting measures to reduce pests and breach of the action threshold ,for example, pests like cockroaches can be monitored using sticky traps and can be placed throughout structures and can be used on a regular schedule for example monthly (Medeiros et al, 2013).

Knowledge of beneficial insects and their prey preferences exist but it is limited in the sense that the impact of their infestation remains undetermined and there is also no predator to prey ratios available to guide in making decisions on how to effectively manage pests. It is possible to monitor the presence and activity of some beneficial species. Especially large predators and parasites. Beneficial insect species that are small, nocturnal and those dwelling below-ground can be difficult to monitor. There different indicators of key beneficial insect activity.

5.5 SUITABLE METHODS FOR MONITORING INSECT POPULATIONS

Insects ecology must be observed by sampling and observing the largest number of insets at the site. The pest should be sampled from the areas where they are in abundance and where they make their first appearance. The main objective of carrying out sampling is to make observations of as many functional groups as possible so that a representative sample can be assessed. When monitoring areas for pests, it should be done during the growing seasons and the monitoring should be done to identify the first sign of an outbreak.

The frequency of monitoring insects depends on insect biology, the nature of the crop when it possesses a low threshold of damage then more intense monitoring will be required. It should also be done on a regular basis for example monitoring can be done on a weekly basis when a pest reaches the borderline to becoming a threat. The size of the area to be monitored will depend on the area of interest, the size should provide a good enough representation of the area. It will also depend on the available resources. The field should also be investigated in a manner that is predetermined it can use a v, u or z shape. It is important to records that are important for making decisions. If a monitoring form is developed, it should be able to provide information on insects that can cause harm or those that are of benefit. The records should also contain an identification in relation to the field plus the date of the survey. Information on the method used in sampling should also exist. The units of the sample should also be noted and recorded (Barbour, 2016).

A vacuum is usually used in the sampling of insect ecology. Insects can be monitored by luring the insect into traps and chemicals can be used as bait, for examples, those they use to communicate such as pheromones. Traps can be used in detecting an insect for example in a field. They are useful in the monitoring of insect populations in order to determine the need for control or timing of control practices. Mass trapping is more effective when the population of the targeted pest is low and if migration into the area is at a minimum. Insect traps consist of a visual attractant such as color shapes or light or one with a chemical attractant to attract the insect being targeted together with a container to trap the insect once it reaches the trap. Most traps make use of glue and sometimes affect the mobility of insects by having a funnel structure to prevent their escape. The methods will have varying performance and some of them will be biased. Some of the sampling methods provide absolute estimates while others provide relative estimates. This will depend on the method whether it is trapping or netting (Barbour, 2016).

5.5.1 Methods of Monitoring Insects and Sampling

Traps: The use of traps can serve the purpose of sampling mobile insects. The traps are left in the locations for a given time period and later the insects in the traps are counted. The traps should be attractive in order to lure the insects that can either be due color or the use of chemicals such as food. The traps can also be passive and capture insects by accident. Traps can be used in sampling predatory and parasitic insects, natural enemies, beneficial insects and epigeic arthropods:

- Colored sticky traps or white traps-They are instrumental for monitoring insects in their adult stage in the field. Yellow sticky traps and water traps are used to monitor adult leaf miners, whiteflies, aphids and thrips among other pests. White and blue attracts thrips. Beneficial insects and other insects are attracted by the yellow strips and therefore should be used if necessary.

They can be made at home with ease. Water traps are used to trap aphid thrips and immature leafminer that move to the soil to allow their transformation into a pupa. Sticky and water traps need checking from time to time. The insects that are caught in the traps afterward should be identified and their number determined and the traps serviced. Stick boards have to be replaced similar to water once the spaces are occupied by insects, dust or debris to ensure their effectiveness is maintained. Other than the use of traps other monitoring method, include the use of visual counts. The advantages of using this method are the ease in deploying the trap, collecting the sample and checking. The disadvantage is that it can be blown by the wind and it requires constant checking and it also requires physical counting of the collected sample.

Figure 5.1: A yellow sticky trap.

Source: https://conversations.echocommunity.org/t/comparison-of-three-insect-monitoring-traps/232

- Bait traps-Used to attract insects making the use of bait for example rotten cheese. They require continuous maintenance in order for them to remain attractive. They may also attract large animals which can destroy the traps while trying to feed on the bait. Some of the attractants used as bait have a chemical nature. They include benzyl acetate, eugenol, vanillin and cineole. Evidence points out that the specimen collected is not in large numbers compared to using bait and netting combined.

- Pan traps- In this case, a bowl is used and contains a killing agent within it such as water or some liquid form. The traps are placed in the area in the morning and collected in the afternoon or even up to twenty hours. Most of the time they are placed on the ground and are sometimes elevated. This method has been used in the collection of bees. Bowls with a number of diverse colors have been used to collect bees. Traps are usually arranged in a transect and regularly spaced to facilitate location ease. Inter-trap spaces can fall between three to five meters.
- Van traps-The methods consist of using two colored plastics cross vanes with a container to collect the insects. After coming into contact with the vane, they fall into the container. They are expensive in comparison to bowl traps, have less ease in transportation and require hanging support. They are not commonly used like bowl traps and lack studies with optimal trap number and spacing but there is evidence that they collect a large number compared to bowl traps. They have been used in the assessment of diversity, abundance and composition of species. They have also been considered less labor intensive in relation to the collection period which is done weekly due to their low rates of evaporation compared to bowl traps.
- Pheromone traps-Contain chemicals which are used to attract males sexually for some species in order to attract the males into the traps. A scent lure may also smell like food. In their natural environment, insects release odor which acts as a signal for mating this applies to the females or attracts other insects from the same species that is aggregation pheromones. The scents can be reproduced in labs. The product is then placed on rubber caps or other devices and it is continuously released over a period of time attracting the insects to the collection device. Several pheromone traps have been developed to monitor insects and also for mass trapping of African bollworms, cutworms, fruit flies and are used widely. Pheromones are not however available in east African traps and are therefore imported but are not also available to small growers. Pheromones are useful for trapping large populations of

pests. In the case of massive trapping, traps are usually placed in different places around the area being monitored to ensure enough of the insects are trapped to reduce the population by a significant amount in order to limit the damage they cause. When aggregation pheromones are used in attracting adult beetles both males and females, the traps lead to the reduction of feeding damage caused by the insects and reducing reproduction by preventing them from laying eggs. When sex pheromones are used, the goal is to capture males before the mating process takes place in the case of moths. This method has become increasingly useful when the key species are known and have developed synthetic chemicals to fulfill this purpose. The traps must be checked and changed from time to time. An example is the wing trap which makes use of pheromones used to trap the grape root borer. A bucket trap that uses both color and pheromones. It is reusable and easy to deploy, the initial cost of purchasing it is high.

Figure 5.2: Pheromone traps used in luring insets.

Source: https://www.treefruit.com.au/media/k2/items/cache/9b2c4b44fb86522964124ed80d03c5e8_XL.jpg

- Light traps-They are used to attract nocturnal insects that fly at night and captures them in a container below the light source. They can be used to capture moths, armyworms, cutworms, stem borers and others. Such traps will work more effectively if they are laid just after the adult moth begins emerging but before they start to lay their eggs. One of the disadvantages is that they attract a wide array of insects. Some of the insects attracted might not even be pests. In addition to this, a number of insects that are attracted to the area do not actually fly into the trap but stay close to the trap increasing the total population of insects in the locality.

- Malaise and window traps-They are used to trap flying insects. They are sensitive to wind and rain and most models are difficult to transport in numbers. In open terrains. They are easily visible and they may suffer from a relatively high sabotage frequency. They require continuous maintenance in order for them to remain functional. They are more passive in comparison to other traps and they tend to collect a sample that is representative of the species that exist in an area. They may also collect species that are not targeted which is a disadvantage unless it is used scientifically. They are labor- intensive as a lot of time is taken in counting the numbers due to abundance in the catch. Other disadvantages include the cost per trap, time use in setting up and the number of traps that can be deployed
- Pitfall traps-These are small containers buried so that their upper edges are just below ground surface level. Ground insects for example beetles fall into the trap and are captured in the conserving fluid within the trap. To prevent flooding, a roof is sometimes used to cover the trap and therefore the contents will be preserved for a period from days to weeks in the event of a flood.

Figure 5.3: A pit fall trap filled with molasses.

Source: https://conversations.echocommunity.org/t/comparison-of-three-insect-monitoring-traps/232

Transects: In this case, an expert makes a predetermined tour on the site and makes a report on the species that have been observed. This method will yield results in different results since it depends on the individual who is carrying out the survey and their knowledge and the results can ,therefore, be hard to standard. This method is subject to bias.

Assessment of herbivores: It is done by direct frass measurement or by directly counting insect larvae. Surveys on the larvae should be done at the right phonological age on the species of insects. There can be difficulty in the transformation of absolute densities. Frass drop management gives direct values related to the consumption of host foliage by insects. It will be necessary to refine the correction factors due to leaching by rainfall. During the growing season, it requires monitoring at intervals of between seven to ten days making this method labor -intensive.

Remote sensing: Can be a tool used to monitor mass occurrences and outbreak of species. Problems sometimes may occur in obtaining pictures not blocked by clouds and sufficient resolution for the period when the larvae are defoliating the trees.

Electronic trap for the automated monitoring of insect populations: One of the frequent and reliable methods for monitoring insects is the use of pheromone traps but at the same time it consumes a lot of time as it requires constant inspection of the traps. In addressing this problem, a low -cost sensor system with the capability of counting species targeted automatically was developed and it is commonly referred to as the z trap. This technology provides integrated pest management programs that are economic and also friendly to the environment since the operations will require less labor in capturing insects. This trap detects and quantifies insect numbers captured by the traps and this information is sent to the farmer through a wireless connection. This technology will also be integrated with pheromone traps and it will lead to a reduction in the use of pesticides. The technology has been used successfully in the collection of data on the codling moth and leaf rollers in apple orchards. The z trap is the hardware itself but it is paired with software that is used in the presentation of data to growers or other individuals making use of it. For each area, five to ten traps can be used.

Figure 5.4: Automated monitoring machine for insect population, z trap.

Source: https://www.slideshare.net/CASCrop/automatic-monitoring-of-insect-populations

Netting: The netting of insects takes two forms that are targeted netting and aerial netting. Targeted netting allows the collection of a particular species within its locality. Aerial nets are made with a light mesh fabric while sweep nets are made with a cloth that is sturdy to allow sweeping through vegetation to collect the insects, sweeping is less precise than netting but it can still provide data on the fauna that is associated with a particular area. Standardizing can be done by controlling time and the period of sampling. Active netting can consume a lot of time, takes less time and variations between those collecting can present variations in collected samples. In some cases, targeted netting has been combined with bait traps for example in collecting samples of orchid bees. Sweep netting involves netting that is continuous for a continuous period. Allows the collection of species that are not collected by targeted netting. Sweeping netting has been used together with bowl traps. For example, to assess *Apis* bees in canola fields.

Direct counts: In some cases, beneficial insects can be sampled with the use of direct counts on vegetation. It is suitable for insects that are generally inactive since accurate counts are difficult to obtain for insects that move fast and are active. Natural enemies on vegetation or crops can be counted visually. It is time-consuming depending on the visibility of the insects. For those that are active, they can be counted after the removal of vegetation with the use of a suction sampler.

Suction samplers: It makes use of a motor for the devices that use suction. They collect arthropods from vegetation. They can be used in the collection of epigeic arthropods. Most of the time they have been applied in grasslands. The efficiency can be specific to the taxonomic group. Their function varies according to year, season, area and the structure of vegetative matter. It can also be affected by factors in the environment.

Figure 5.5: Suction samplers used in sampling insects.

Source: https://www.researchgate.net/figure/Sampling-invertebrates-at-Needingworth-using-a-Vortis-suction-sampler_fig4_269409248

5.5.2 Sampling Beneficial Insects

They are made up of a wide variety of insects and they are sampled using a variety of methods. Most of which have both advantages and limitations. The sampling methods will use different results but they can complement each other. Consistency is necessary for the results to be comparable. Using the combined methods can increase completeness in comparison to using a single methodology. Research is required to for the development of new methods for sampling for insects that are beneficial and comparing the different methods.

Sampling methods can be lethal to insects that are beneficial, their abundance and diversity. Most studies that have done are on sampling effects on vertebrates. In the case of invertebrates, there is no evidence of negative long term impacts on insect communities but caution should be taken to avoid over- sampling. Reliable sampling and methods from monitoring for insects that are of benefit are important in the assessment of the effects of disturbance to the environment and effectiveness of efforts to do with conservation.

Bowl and blue vane traps can be used in sampling bees, wasps and flies to determine relative abundance. Malaise traps for flying insects targeted netting for orchid bees and other flying insects. Sweep netting and direct counts for insects on vegetation. Suction samplers for those found in the litter. Pitfall traps, soil flooding, cryptozoan boards and emergence traps apply to epigeic arthropods. Trap nests are used to trap nesting bees and wasps. Stick traps are used for parasitic Hymenoptera

5.5.3 Monitoring Insects for Conservation Purposes

Butterflies are one of the major groups of insect which are experiencing a decline in their population. They need to be monitored in order to support measures for conservation. Monitoring efforts can make use of citizen scientists. Different methods can be used including malaise traps, pheromone traps, light traps, time counts and transects. The data we get from this is used to keep track of population trends of insects in order to take note of whether the species is increasing or decreasing we need to identify the name and the species of the insect and depending on the numbers one can dedicate efforts to conservation. Some common methods used by BMS to monitor butterflies in Britain are the transect counts and the 15-minute count. Difficulty in the use of the transect counts has led to preference of the 15-minute count as it is flexible in its use.

Transect counts: A transect is a fixed route that is formed in an area where insects such as butterflies are recorded weekly for a period of years and make use of some basic rules. The select sites are decided on by the individual making the counts. Other schemes provide suggestions on the areas to record in order to get records to get even lad coverage as well as ensuring an array of species are monitored. The counts can be done individually or in teams.

- Transects cover about 1 km which is divide into sections that correspond to different habitat areas.
- Butterflies or insects can be counted in the presence of adults for example when they are flying regionally as the flight time will differ.
- People walk following the routes transected on a weekly basis during seasons when they fly. If this is impossible, then counting can be done every two weeks or 10 days.
- Butterflies are counted in an invisible box of 2.5 meters on each side with a height of 5 meters.
- The visits should be done in the course of good weather especially when it is warm and sunny.

Figure 5.6: A transect used in monitoring butterflies.

Source: https://butterfly-monitoring.net/bms-methods

15 minute counts: This type of monitoring is to get information on the abundance of insects and in this case, butterflies. It can also apply when monitoring rare species and their behaviors. The method is focused on monitoring insects in certain areas for a fixed period and in this case, the period is fifteen minutes. This type of method can be applied in the case where people know the species they are monitoring. Maintaining the same time will ensure uniformity in the results. These methods have been used by the BMS for butterflies such as *Mellicta athalia and Argynnis adippe* in the United Kingdom.

6

Conserving of Insects Through Regulating the Use of Pesticides

CONTENTS

6.1 INTRODUCTION

Most of the insects have been known to be a nuisance to human beings with some of them being pests. They have led to the destruction of the crops that have been planted by farmers. Some of them even destroy stored foodstuff that have been stored to help sustain the environment. This has involved the use of pesticides that includes weeds. Pesticides are a collective name used for chemicals: herbicides, insecticides, insect's repellents, bactericide and fungicide among others. Most of these pesticides have been used in plant protection as they have been to some extent to be used in the protection of plants from fungi, weeds and insects.

In as much as the use of insecticides has been effective, they have led to the destruction of insects. Insects have been said to be a nuisance but to some extent, they are useful in the ecosystem in the aid in the maintenance of the environment. They are utilized as food. Take the case of insects such as grasshoppers and crickets that have been used as food in a variety. In order to understand the need for conserving insects, we need to look at the importance of insects.

6.2 IMPORTANCE OF INSECTS

There is much importance of insects in the environment more so in biodiversity. This may be viewed in the line of biological control where insects are used in aiding a variety of biological processes such as controlling pollution by breaking down wastes present in the earth's surface reducing the levels of pollution on the land. In biological control, insects are used to feed on potential predators to certain crops minimizing the damages that would have been done on the crops. Some insects also feed on weeds leading to the reduction of the spread of the weeds.

Insects can also be used as food for humans which is a practice that has been conducted in Central and South America among other countries. Through research that has been conducted it has been established that there are more than 1400 species of edible insects. There are various ways in which the insects can be ingested in most cases, it involves them being fried. Insects have also been used for medicinal purposes as they have been used in curing various medical issues such as, "Apitherapy" that is healed through therapy which makes use of venom from bees.

This kind of therapy is said to contain a range of amino acids, proteins and enzymes resulting to the production of a hormone known as cortisol aiding people who deal with bursitis and tendonitis among others. There are also surgical maggots utilized in maggot therapy. In some cultures, when there are stitches needed to close the wounds there are insects. Insects are also useful in pollination for the flowers that are insect- pollinated. The insects carry the pollen and they transport them to the plants that are to undergo fertilization. Insects are used in recycling. For instance, the leaves that fall from trees without the insects, they would have accumulated. Insects eat the leaves thereby breaking them down resulting in the release of nutrients that aid in allowing growth in plants. The insects utilized in this type of feeding are known as Saprophages. There are insects whose diet consists of dead animals and plants.

There are insects whose diet is mainly the tissues of dead animals and they include ants, wasps, flies and beetles. There are those whose diet includes the faces from other living organisms and are known as coprophages. Some of the insects with this diet are the dung flies. Insects are also used in making useful products which include making the cochineal dye, shellac, honey and silk.

All these make the need for these insects. This creates the need to conserve the insect species. Some initiatives have been taken up to ensure the conservation of insects which include them being bred by farmers such as the case of bees. The other involves the minimization of the use of pesticides. The pesticides have contributed to a significant decrease in insects. The pesticides are chemicals that when they enter into the body of the insect's they inhibit growth or they cause death immediately or gradually. In a bid to understand why the uses of pesticides need to be minimized, let us see the next section.

6.3 EFFECTS OF PESTICIDES ON INSECTS

Pesticides when ingested by insects tend to affect their immunity. They, therefore, affect their effect on the agricultural sector. The access to the insecticides is dependent on the mobility of the insects. For instance, like bees, they are well known to pollinate plants and therefore they are more or less likely to be accessible to the pesticides. The pesticides are major affecters to the immunity of the insects. The chemicals are known to suppress the immunity of insects. For instance, the organophosphates as well as the botanicals have been known to affect the immunity as they impact the hemocyte number, differentiation, therefore, affecting the phagocytosis. There is the phenoloxidase cascade ,as well as the toxicology of the skin, has been said to influenced by a number of pesticides (Soloneski, 2014).

Most of the synthetic insecticides are said to increase the oxidative stress and may pose significant implications on the generation of some of the antimicrobial peptides in insects. The pesticides may influence certain behaviors such as grooming making the insects be prone to a variety of diseases. The pesticides also affect the nervous system of the insects. Therefore, it affects the transmission conducted by the neurotransmitters. It may affect the chemicals used in transmitting the information through the use of the synapses known as acetylcholine. There is an enzyme known as cholinesterase that binds with the acetylcholine, which allows the muscles to rest. When the pesticides are introduced they keep the cholinesterase from binding with the acetylcholine. This leads to the excessive stimulation in the muscles that may result in the paralyzation to eventual death.

The lethal introduction of these pesticides on the insects to high levels leads to the significant death of the insects. The pesticides may have both temporary and permanent effects on the insects. In the case of the permanent effect, it may the reproduction process it may affect the hormones used in the bodies of the insects. The pesticides may affect the environments of the insects (Soloneski, 2014).

The insects have a variety of effects on the environment beyond the insects. Recent research has indicated that there has been excessive use of pesticides which has summed up to high levels of the destruction in the surrounding in the soil, water and air among other elements of the environments. There are a variety of pesticides that have been used in the harming of insects minimizing the damages that have been caused by insects. It is important to know the pesticides so as to know the effect that may be caused by them so that we can identify a better method that can be used instead of the pesticides or the levels to which the pesticides can be used so that it cannot have server impacts on the surrounding (Soloneski, 2014).

6.4 TYPES OF PESTICIDES

Various insects are used in agriculture. In the case of insects, the pesticides that are utilized are the insecticides. There are other pesticides include in the herbicides that are used in the protection of plants , bactericides used on bacteria, rodenticides used on rodents, fungicides used on fungi and the larvicides used in the case of larvae.

As we are focusing on insects, we shall focus on insecticides. The insecticides are utilized depending on the habits of the insects. The habitat will also affect how the insecticides will be administered to the insects. There are four modes in which insecticides can be administered to the insects. Either in solid, liquid, aerosol or granule forms. A better illustration is with the case of the flying insects; the aerosol insecticides are the best mode to combat these insects. Also, the use of fogs may be effective to some extent. In the case of the crawling insects, the appropriate method may be the use of surface powders, sprays or granules. The mode of administration of the insecticides will be dependent on the type of entry that will be utilized when the insecticide is administered to the insects.

The insects work on the principle that involves them entering the body of the insect and affect certain parts of their bodies ,therefore, taking the form of poison to those areas. There are three modes in which the insecticides can get into the body of the insect. They include respiratory entry, oral entry or dermal entry (Soloneski, 2014).

6.4.1 Skin Entry (Dermal)

This mode of entry involves the insecticides entering the body through the skin hence the name dermal. In the case of the insects, their skins are commonly referred to as cuticles. The insecticides utilized in this case are known as contact poisons. There are certain scenarios where the dermal entry is used: the first scenario is when the aerosol spray droplets hit the insect. The other case involves the insects walking over and thereby come into contact with powder or granules forms of the insecticides.

6.4.2 Oral Entry

This mode of entry involves the insecticides entering the body through the mouth hence the name oral. The insecticide is ingested when the insect eats the drug. The insecticides used in the case of oral entry are the ingested poisons. Some of the insecticides that may be ingested by the insects include a poisonous bait that includes poisons being added to food that is to be ingested by the insects. These poisons may be ingested as the insect grooms itself after the poison comes into contact with its body.

6.4.3 Respiratory Entry

This involves the insecticides breathing in the insecticides. The insecticides used in this case are known as inhaled poisons. The insects different from other animals do not breathe with the nose but the breathing occurs through the spiracles.

The other pesticides may not directly affect the insects but they affect the environment of thc insects which includes their foods. For instance, in the case of the herbicides, they are chemicals that kill plants by either killing the parts that the chemical comes into contact with or killing the parts of the plants that absorb: the herbicides which can be the leaves, stems or the roots. When these parts are killed, they deny the insect's food. Though some of the insects feed on dead organisms the parts that have been killed with the herbicides may affect the insects when they ingest the foodstuff having the chemicals in them. They are more or less like human beings when they ingest chemical compositions (Soloneski, 2014).

When using pesticides there is a pesticide treatment that is used so as to ensure that the use of the chemical is not done extremely causing an increase in pollution. The use of this program involves the use of one or more pesticides. The program clearly defines that the pesticides should be used only when there is need such as in the case where the pest problem has been identified which include extensive infestations of certain insects like cockroaches in homes of human beings. If a pest control method is to be initiated-there options that have to be weighed and the appropriated and the instructions correctly followed. However, though this program has to be followed, it has not been the case. Due to the lack of knowledge by most of the farmers among other users, they have made excessive uses of the pesticides thereby causing pollution to the environment and also causing server damages to the populations of the insects on the land.

This has led to the need for regulations being being made with regards to the utilization of pesticides. There have been minimizations that have been made when using pesticides so as to reduce the impacts they have caused on the insects (Marrs& Ballantyne, 2004).

6.5 TYPES OF INSECTICIDES

In most cases, the pesticides that are mostly utilized more so around homes are insecticides that are used in killing them as they are a nuisance. There are a variety of insecticides that include synthetic pyrethroids, insect growth regulation, inorganic pyrethrins, organophosphorus insecticides and organic insecticides. Insecticides that are of mineral origin are the inorganic insecticides and may be composed of substances that include arsenic trioxide powder and the boric acid. This type of insecticides is not commonly used as they are said to be toxic to most of the mammals and they are also non-biodegradable. For these types of insecticides, the action experienced is stomach poisoning. These are used as baits. They are administered in solid form (Jack &Rechcigl, 2016).

There are the organic (natural) pyrethrins, these insecticides made from a specific plant. The common one is pyrethrin, which is obtained from the chrysanthemum flower. The use of pyrethrins is advantageous when used as an insecticide as they are of a broad spectrum. They are also of reduced toxicity in human beings, there is a fast response as it is quickly knockeddown and the other advantage is the fact that it is biodegradable. The major disadvantage of the pyrethrins is that the residual action may be absent or slightly experienced. In the case of the synthetic pyrethroids, the insecticides have been made chemically so that they can work like man-made pyrethrins. They have low toxicity to humans and are said to be effective for a variety of insects. Some of the synthetic pyrethroids include coopex, permethrin, tetramethrin, bioresmethrin, deltamethrin and bifenthrin among others.

There are also organochlorine insecticides they are insecticides that are made up of chlorine as the major component. Some of them include heptachlor, endosulfan, chlordane, DDT and dieldrin. They can mostly be utilized as oral and contact poison. The insecticides work on the nervous system. These insecticides are no longer used as they have persistence and impacts on the environment. There is only a single organochlorine that is registered in the permit in the agriculture. The others were deregistered for use in the year 1996.

The carbon chemicals are used in the manufacturing of synthetic organic insecticides. Some examples of these insecticides are organophosphorus insecticides. They may entail phosphorous. Some of them are dichlorvos, chlorpyrifos, temephos, diazinin and malathion. The insecticides under this category are highly toxic to animals. The use of these pesticides is regulated to minimize the interaction with non-targeted insects. These types of insecticides are known to break down in the environment quickly compared to the organochlorines. There some of them that tend to be active for months to years. These insects are said to have a strong chemical odor (Soloneski, 2014).

There are carbamate insecticides that are said to be relatively unstable. They are broken down in the surrounding in a given duration of time. Propoxus is a commonly used carbamate. In chemical products such as Baygon, these components are highly effective. They are used as oral or contact poisons. They can be utilized as bait or surface sprays baits. They are used in the management of pests found in homes of human beings. The use of these pesticides has led to the extinction of insects and thereby affecting the ecosystem. There have been a variety of mechanisms that have been taken up to minimize the effect. One of the mechanisms has involved taking other alternatives to control insects (Soloneski, 2014).

6.6 ALTERNATIVES TO PESTICIDES

This topic is usually discussed in organic farming. As the pesticides are known to deteriorate the environment when used continually. Most farmers have taken up organic farming. Three agricultural alternatives have been taken up. They include Biocontrol which involves using the pest's natural enemy to keep them off from the farms or homes. This has involved the use of extensive research being done so that the predators do not cause damage to the intended use. This implies that the predators should focus more on eliminating the pests. This method is not newly invented and is been taken up by other countries.

There is the polyculture which is also known as companion planting. This method involves the cultivation of different types of crops in a particular field instead of planting one species of crops in a given field. This is the same concept that is used in polyculture it takes the form of planting companion plants along with crops. This is a strategy that has been taken up to support the crop. This method is mostly utilized in plants that play the role of repellants of insects to certain crop pests. A good illustration is when the tomato plants are planted together with plants such as cabbage. Tomatoes in this case act as repellents to pests such as diamond -backed moth larvae. Good examples of natural repellents to mosquitoes and flies are the basil tomatoes (Jack &Rechcigl, 2016).

The third method involves the use of natural barriers and predators. This is better illustrated by the cultivation of tussock grass by the UK farmers. They cut across the center of the land this gives a habitat to the spiders as well as beetles thereby protecting the other crops from pests such as aphids. Using other methods has offered a variety in scalability in matters of farming and production in agriculture. With the use of the agricultural method of removing pests, it is a method that will ensure that the important pests are safeguarded. This ensures that the ecosystem is well maintained. The other method involves the regulations involved with using pesticides. There are official bodies that are in charge of regulating the use of pesticides (Jack &Rechcigl, 2016).

6.7 REGULATION OF PESTICIDES

The use of pesticides has been known to significantly work in minimizing the number of insects on the land. As a result, it has led to the destruction of the environment as well as the ecosystem. Over the years it has been noted that there has been excessive use of pesticides by farmers and other users. In most cases, the regulations that have been made concerning the use of pesticides are managed by the EPA. There have been acts that have been made concerning the use of pesticides in different areas. Some of the acts include the FIFRA. The FIFRA requires all chemical products sold or dispatched to different areas that should be registered. In the FIFRA there are different lists which the pesticides can be registered in. They include the following:

Federal registration actions in the United States is controlled by EPA. They are in charge of the registration of pesticides. All the pesticides that have been manufactured are to be tested while under development. This is all included in the Experiment Use Permits (Marrs& Ballantyne, 2004).

The registration of new pesticides by states is facilitated by the State Specific Registration. This allows insects to be used for non-specific purposes. It can also be used for products registered federally for other uses in the case where it has been established there is a certain local need and endurance or another clearance through the FFDCA. The governments in most cases are the ones that are in charge of picking pesticides that will be used in their countries. The EPA is in charge of the regulation of pesticides that are listed within the Federal insecticides, fungicide and Rodenticide Act. They are therefore tasked with ensuring that the manufactured products will not cause harm to the surrounding and humans as well.

There is also a designated restriction so that some people are given the mandate to purchase pesticides. In the United States, there is a controlled usage of pesticides. These are restriction involves the pesticides not being easily accessed by the general public. It, therefore, restricts a product that is a pesticide. It is therefore applied by a certified pesticide applicator. The certified applicator is the one who conducts the supervision. This implies that one has to have a license in order to purchase these pesticides and also apply the pesticides. There are certification programs that are given by the federal governments and also companies that vary in the countries. The EPA manages. There pesticides that are underlined under restricted use due to a variety of reasons. Some of the reasons may include the potential or the history of groundwater contamination. In the case of pesticides, the atrazine, it is the herbicide that is highly restricted since 2017 (Marrs& Ballantyne, 2004).

There are a variety of requirements that are needed when dealing with pesticides before one is given the license to use pesticides. There is the Worker Protection Standard which establishes some of the requirements that are needed so that one is given a proper license so that they can be able to purchase and use the pesticides while under restrictions. The Worker Protection Standard needs the individuals to be aware of the hazards. The supervisors need to uphold the records of applications for a given duration of time. The number of years is determined by federal laws. The records need to indicate the location, date and the kind of pesticide that will be utilized.

The local government is notified by the pest control supervisors. This part of the government is concerned with the quality of air and are notified so as to ensure that certain rules regarding air quality are met. This is with regard to public health and safety threats with regards to controlled use of pesticides and they are applications of these pesticides. The different kinds of pesticide applicators when using pesticides do so in line with the laws and regulations that have been set. The laws offer guidelines used in the registration of pesticides. There is also a system for classifying pesticides for the general or restricted use of pesticides. They can be used in the classification of pesticides. They can be used either as a restricted -use or general -purpose. They are used as guidelines in labeling pesticides. They are also utilized in the training and the certifying of systems used by farmers who tend to make use of restricted pesticides (Marrs& Ballantyne, 2004).

6.7.1 FIFRA

This is the primary law that is in charge of labeling, registration, use and selling of pesticides. It is needed that the states should follow and meet the standards listed in FIFRA. Numerous amendments have been made to the FIFRA. Some of the amendments include that of the requirement that pesticides should be categorized as either restricted use or general use. A pesticide that has been used for general use in FIFRA is one that will not result in the cause server implications to the surrounding when it is utilized with regards to the labeling. These kinds of pesticides are usually available to the general public without them needing to have a license. In the case of the restricted use pesticides are those that are defined in FIFRA to be pesticides that will generally lead to unreasonable diverse impacts in the surrounding as well as the individual applying the pesticide (Bergeson, 2000).

Two conditions may govern the application of pesticides. One of them involves the process being undertaken with the presence of a licensed applicator supervising it. The other condition is that the laws listed by the EPA should be followed. The 1972 amendment of the FIFRA established two categories of applicators. They include private or commercial applicators. Licensed applicators also are known as private applicators. This applicator will oversee the use of any restricted use pesticides. This is all done to generate any agricultural products on different kinds of property. There is in place, the Montana regulation which allows the use of pesticides for general purposes for a direct or a closely located landowner. There are two ways in which the private farm applicators become licensed. This includes undergoing a pesticide training program and once they are done with a test after finishing the training. The other method includes one doing a farmer applicator test even though they have not undergone the required training. In the event where a score of at least 70% one is able to get a certificate (Bergeson, 2000).

In the case of a commercial applicator under the definition of the FIFRA, it is a licensed pesticide applicator who makes use or oversees the application of pesticides. In order to become a licensed commercial applicator, they need to have practical knowledge regarding the activities that are undertaken when using pesticides. This is achieved by taking a test that will be graded. In the case where one gets a grade of 80% and above then it is a pass and is qualified to use restricted pesticides. For those who get a 70% -80% then they qualify to handle general use products.

In 1996, FIFRA was amended by the Food Quality Protection Act. The changes that have been added include determining a system used in revising the registration of pesticides and tolerance on the basis of a given cycle. There is the permission needed by the EPA in order to remove pesticide registration once there is a condition said to be an emergency and the conditions developed by the EPA to allow minimization in the hazardous threats posed by the pesticides as well as the fee imposed in the registration of the pesticides. The EPA gives the laws in the FIFRA through this act outlines that the countries have to be primarily in charge of taking up responsibilities if they are able to prove to the EPA that they have put in place the needed laws and mechanisms. EPA incorporated the use of Montana. They shoulder the role of training and testing the private as well as commercial applicators. When in a given area the mandate of playing the role of a supervisor is taken up by the EPA. However, they still monitor and regulate the registration of pesticides as well as the monitoring of pesticide producers (Bergeson, 2000).

There are civil or criminal penalties that have been listed in the FIFRA that are to evaluate for violations of the provision in the Act. Some of the violations that are included while under the FIFRA include the selling and restriction of the use of pesticides by an individual who is not licensed to buy these products. The other penalty also includes a lack of consistency with the labeling and the pesticide in question. There was the cancellation of certain pesticides which over the years have been sold to people against the laws in the FIFRA. It has been noted that some of the applicators were not keeping the registration records which would have been utilized in the inspection process. Whenever we talk about pesticide regulation, we can talk about the Montana Pesticides Act. The Act was put in place to handle a variety of roles which include the registration of the pesticides as well as the giving of the pesticide certificates.

6.7.2 Pesticide Registration

This requires that pesticides utilized in Montana need registration within the EPA. The Montana Department of Agriculture is the body in charge of registration of pesticides in Montana as outlined in the EPA. This body is also in charge of sampling, inspecting and conducting evaluations on pesticides that are dispatched in Montana to ascertain that people adhere to the Pesticide Act in Montana. The registration of the pesticides should be in line with the FIFRA. There is the licensing Provision which requires the pesticide applicators to be individuals who apply the pesticides on land such as the Suburban or on their vegetation. Some individuals are paid to apply the pesticides. These appliers of pesticides are called commercial applicators. By law, it is required that an individual who makes use of restricted -use pesticides should be owners of a private applicator certificate. When dealing with the general -use pesticides, these groups of individuals do not need any form of licensing for them to purchase it. Some regulations are in place that are considered to use pesticides (Marrs& Ballantyne, 2004).

6.7.3 Regulations in Pesticide Use

For a pesticide to be chosen they usually undergo a number of tests on the basis of the effectiveness and safety of the pesticide. In the FIFRA there are a number of laws in it. They include the following:

There are a number of evaluations that are done to evaluate the effectiveness of the pesticides which include the potential risks of the material, the use and the required dosage. All this is done after the collection of data. The labeling instructions on the packages are listed to increase the productivity of the pesticide ensuring that the applicator is protected. It the protection of the environment as well as the consumers. The general public can access a given number of pesticides. A huge number of pesticides are said to be highly dangerous and cannot be used for general purposes. A system has been set up by FIFRA of testing and licensing of pesticides at the private stage and the commercial stage for individuals who want to buy the restricted pesticides. There is also monitoring the controlled acaricide (Wheeler, 2002).

There is a process used in reviewing the categories of acaricides which include conventional pesticides, biopesticides and antimicrobials. The classifications have a common application process; they may need varying data requirements. The categories of the pesticides may affect assessment procedures which may run for a given duration of time. When the insecticide is recorded in the EPA there may be some considerations made by the state with regards to the registration. Countries may outline an extra set of laws and regulations concerning the registration of pesticides. To some extent, countries may be required to make requests for the yearly utilization reports from users who use these pesticides.

There was an improvement of the Pesticide Registration Improvement Act 2003 changed the authorization of the fees for a given product, they evaluated the procedure used in the gathering of the maintenance fees. They came to the conclusion that an evaluation procedure in assessing the pesticides. PRIA plays the role to make sure that there is easy enforcement of pesticide laws to the people who utilize it.

In the importation and exportation of acaricides, there are things that are underlined. They may be required to have total Notice of Arrival by the customs in the states and the protection at the borders. In the case where the NOA has not been finalized then the product will not qualify to make it across the customs. Under the Notice of Arrival, there is a list that identifies the commodity, the quantity in the parcel, scheduled day for arrival and location of inspection. A set of laws that are covered under this. They include that importation and cxportation should comply with standards that have been set in the United States with insecticide laws (Wheeler, 2002).

The other rule is that it should not be tampered with or volatile and proper labeling is required. It is also required is that the commodity should be generated in an EPA record established yearly. In the case of the pesticides that are to be exported to different locations in the globe may not require recording of the conditions under given terms. Some of the terms may include: the purchase having to give a statement saying that the commodity has not been registered and may not be put for the soil in the United States. The pesticides should have a label indicating that it is not registered for usage while in the U.S.

The requirements that have been listed on the label should be met. The instructions should be written both in English and the receiving state. The pesticides need to be in line with requirements that have been listed in the EPA. In the year 2016 on 22 April, the Inspector general's office announced their intent to being preliminary research to assess the inspections done by the EPA. They also enforce against illegal pesticide imports. The project had the main aim of establishing if the EPA's Federal Insecticide, fungicide and Rodenticide Act utilized an inspection manual is good in dissuading, establishing and holding back unwanted pesticides brought in the country ensuring the protection of the environment (Wheeler, 2002).

6.8 REGISTRARING PESTICIDE PRODUCTS

In the case where there are pesticide products to be registered in the EPA, It should be understood the consideration listed in the law. This is under the section 2(u) of the FIFRA. This is pesticide is said to be as follows: any matter that has been meant to be utilized in the prevention, destruction, and mitigation or repelling of certain insects. It also refers to as a combination or singular materials that are needed and utilized as plant regulators, desiccants or defoliants. In some instances, it includes nitrogen diluters unless in the case the name acaricide does not entail one that is a newly found animal drug. The individual making the application is required to demonstrate that the active ingredients in acaricide, pesticide products or newly proposed recorded acaricides may lead to unmeasurable impacts on the health of the surrounding or that of living beings.

The unmeasurable extreme implications may show possible risks that may be server to human beings or the surroundings that may include economic, environmental or social costs. This may include their values which may be looked into. It may entail the risks in the feed that would have been brought about by the acaricide utilized with food and has no match to the set conditions listed in the Federal Food, Drug and Cosmetic Act (Gunther, 2012).

It is required that the applicant should give scientific data resulting from mixtures of 100 different assessments done listed the EPA directives to evaluate the extent of the different significant effects. There has been some demonstrated interest by the public in matters registration of acaricide research that leads to the arguments over the people's being able to acquire the substance. The competing interests as well as the legislation that addresses the access has been separately covered under the Pesticide regulation in the United States. There are instances where the manufacturers decide to conduct additional work to satisfy requirements for peer-reviewed literature as in the case of the herbicide, cloransulam-methyl. Some fees are imposed with regard to the use of pesticides. The fees may vary from maintenance to others. The fees may be reduced or increased or waived depending on the type of the size of the business, public health of the pesticides or the minor use of pesticides at the discretion of the EPA. The failure to pay the registration fees or the maintenance fees may lead to the cancellation of product registration.

In this section, we are discussing how insects can be conserved through the minimization of the use of pesticides. This implies that there are some species of insects that have ended up being extinct due to the use of pesticides. The endangered species are covered under the Endangered Species Act that protects them. In the safeguarding programs, there are two major objectives. They include the provision of the appropriate safety of species of insects rendered as endangered from acaricides and reduce the effects of the activities on individuals who use the pesticide. In a bid to safeguard the species whose lives are in danger there were some rules that have been implemented. The reliable science being utilized in assessing risks to endangered species. A trail that was made at coming up with modes to prevent issues regarding the stated insect species. Also, there were issues raised with regard to the listed species. The scientists dealing with the given insects were included in the consultations so that decisions can be achieved (Gunther, 2012).

They include: adding an inclusive marker to the acaricide, developing bulletins containing home areas and acaricide usage limitation, dispatching the indicators having the information to individuals who use insecticides and provision of a toll-free contact for people to get information concerning the use of pesticide in bulletins and modes they can acquire the information. When discussing the regulation of pesticides, we look into the regulation of non-pesticides products that do not require registration. This involves the use of the Adjuvants which are chemicals that are included to either improve the productivity or modify the properties of acaricide agents. There are over 200 pesticides registered under the EPA that point to certain additions of a number of adjuvants to pesticide combinations so as to ensure that there is an improvement in efficiency. These kinds of chemicals are identified as other ingredients.

Some of the adjuvants include the following: compatibility agents, acidifying agents, oil surfactants buffering agents, oil surfactants, anti-foam agents, defoaming agents, crop oil concentrates, wetting agents, deposition agents, foam markers, anti-transpirants, feeding stimulants, drift minimization agents, herbicide safeners, suspension agents, adhesive agents, emulsifiers, gelling agents, dispersing agents, synergists, neutralizers, water softeners and spreaders among others. The devices that may be used in killing insects other than the pesticides. These methods however do not require any form of registration. The pesticide intermediaries may be utilized in the making of the insecticides during the manufacturing process. When used, the intermediate pesticides are regulated by the Toxic Substance Control ACT that was established in 1976 (Bergeson, 2000).

In as much as it would require a collective responsibility in minimizing the use of pesticides to protect the ecosystems and endangered species such as insects, most of the work is taken up by the government. The government is able to achieve this through the use of the EPA. In the United States, the congress gave the mandate to the EPA to make the laws and regulations concerning the use of the insecticides. The EPA therefore along with the Country Agencies work to review the data on acaricide safety as well as the registration of the pesticide products, education of the professional applicators, monitoring compliance to the laws and the investigation of the pesticide problems. The other areas that are brought under consideration when dealing with matters pesticides are the pollinator protection, endangered species, employee safety, decreasing acaricide drift and reporting acaricide cases.

They also give all the guidance that is needed when transporting acaricides between the states. The EPA can handle most regulatory issues present when dealing with the pesticide programs that have been initiated. The EPA also conducts an evaluation of the new pesticides and those that are being proposed to be used in the environment. It also evaluates the possibility of dire situations guaranteeing short-term approvals for specific pesticide and conducts an often review on the recent investigations in conjunction with the safety of initially used acaricides.They are in a general body that is in charge of pesticide regulation and provides support to state as well as the regional EPA programs that have been designed to protect and train the pesticide applicators (Wheeler, 2002).

6.9 PEST CONTROL METHODS

There are three methods that can be used in and lead to pest control. These methods are used as a means to minimize the use of pesticides. These methods can be used depending on the type of environment. For instance, in homes, there is the use of the Hygienic pest control methods that includes one maintaining a perfectly clean home, which will reduce the existence of the infestation. It is commonly known that pests tend to live in areas where there is an abundant food source and the environment will allow them to breed. Some of the activities that are listed in this method include cleaning all surfaces after eating, keeping the worktops, tables, floor and cupboards clean, sticking to a regular cleaning schedule, storing food in airtight containers, installing flying screens windows and checking the internal and external walls for holes and cracks and seal those that are found. This will ensure that the insects are kept from the home without harming them in any way.

The other method is the biological pest control which is said to be the natural method as it involves the predators provided by nature to get rid of the unwanted pests. It makes use of the hierarchy provided by nature. This will involve the introduction of a predator who by nature preys on a given pest that is present in the environment. This method is used in the case where a population of insects is to be destroyed without the use of chemicals.

Among the common method involves the use of ladybugs in the control of aphid infestation. There is the use of microorganisms as a method of pest control. This method is utilized as nature's method of controlling pest that is utilized by farmers. It works in a similar manner to using natural predators. This method is a preemptive method. There are useful microorganisms that have a symbiotic relationship with the plants. The relationship helps the plant in that there will be an active protect the plants through the keeping off of pests as well as the destruction of harmful fungi and bacteria.

There are physical methods used for controlling pests such as trapping, killing and the removal of insects as well as rodent pests. Some of the methods include the elimination of the breeding grounds. The best method of destroying these breeding grounds is to ensure high levels of cleanliness. Also ensuring that the areas around the given area are not damp. There is the use of the poison bait which is said to be highly effective. It makes use of pest -specific poisons. Though this method involves the killing of insects, it however ensures that the insect that is to be removed is the only one that is eradicated. It works by placing in the areas that demonstrate pest activity and will be eaten directly or be taken back to the nest. This method needs one to be careful as children need to be kept off from them.

There is the field burning. This method has been used ever since ages. The fire burns the debris as well as the soil layers at the top with the aim of controlling the number of the pests. The procedure has not been championed for in recent years as the burning destroys the important insects along with the pests and may also incur bad impacts on the earth's surface.

There is also trap cropping which is a method that utilizing the cultivation of decaying plants in which insects have been attracted to. The procedure aims to trick pests in decaying land so as to cause removal to be much easy. This method has been deemed not to be effective in large tracts as the pests tend to make their way back to the main crop. This method may be effective in small homesteads. It can also be utilized as a method of keeping pests away from a personal vegetable garden.

There is the use of traps which is a physical method and is the most recognizable pest control method. There are a variety of available traps. For instance, there is the simple flypaper to more complex baits, which are used in dealing with insects. This method should be focused on areas where there is evidence of insect activity. This method is disadvantaged in that it will require one to constantly check on the traps so as to remove any creature that may have been caught. There are benefits to using natural methods.

6.9.1 Benefits of Natural Methods

These are natural methods that do not have any potentially harmful chemicals therefore there are no possible side effects of using these methods. Once the natural insect predator becomes established and the plants will have long-lasting protection. By using this kind of method, the insects will not develop any resistance as there are no chemicals required and therefore this method will be effective in the long run. These methods that are used are usually environment friendly.

6.9.2 Disadvantages of Natural Methods

These methods may prove to be difficult to kill the entire population of the pests. The pest populations may move or spread over a larger area. Also, the use of natural predators and microorganisms is more of a reactive measure than a preventive measure implying that there is more time for the infestation to grow.

These methods have been used in place of the chemicals. The challenge has been the effectiveness of these methods that have led to people leaning to using pesticides. These methods are advantageous for use compared to chemical methods is that by using the chemical methods, there is a high likelihood of the insects developing a resistance against the chemical. When insects become resistant, the pesticide undergoes an evaluation and there are pesticides that are made to be much stronger than the previous once and therefore may cause adverse to the environment as well as the health of the human beings.

With the use of pesticides, they can be minimized in the environment by following the instructions that have been listed. It has been noted that some of the farmers and users use excessive pesticides because they do not follow the instructions and may end up using more than what is required. Some of the guidelines that are listed when using pesticides include the following: one should always read the instructions that have been outlined in the packaging of the pesticides. One is also required to choose the right pesticide for the required job and the pesticide should be applied only to the areas that are affected. One should use the amount of pesticide that is required. One should try to spray the pesticide when there is little or no wind.

When spraying, one should ensure that other people or animals are not nearby. In the case where there is the treatment in the house, one should open the windows and the doors and afterward the lingering chemical smells will be removed. As per the instructions, one should not allow people or animals to the sprayed area as it could lead to possible contamination. The major reason for the minimization of the pesticides is to protect the pollinators. Insects play a major role in the process of pollination more so in the insect -pollinated crops. This is because they are used in ferrying the pollen grains from one plant to another thereby allowing fertilization and hence the production of food in plants. The insecticides ,therefore, have to be minimized so that the process of fertilization will be guaranteed. The use of chemicals such as neonicotinoid on flowering plants such as the linden plants have proven to cause concern for the health of the pollinators.

As a measure to minimize the use of pesticides, most of the municipalities have opted to remove the use of the neonicotinoids in various activities such as pasture management. They have opted to use other scientific methods that will ensure that the effects of pesticides are minimized and will not lead to significant damages to the insects as well as the economy. When these chemicals are in use some practices are undertaken to ensure that they do not destroy the insects or the environment. Some of them include pre-application considerations which include mowing the treatments, timing the application of the insecticide, insecticide formulation and choosing of insecticide class.

Timing the application of the insecticide involves applying the insecticide at the right time. Timing involves looking at aspects such as the environmental conditions in matters and the season of the environment. When all the factors are accurate then it will be suitable to apply the insecticide. There are also post application measures that are undertaken. Some of the methods include irrigation which is used in reducing the remains from pollinators foraging areas of the medicated crops and they also alter and reduce the active component concentration. By so doing it reduces the hazards that may have been imposed on the pollinating insects.

Therefore, some habits are created to ensure that there is the protection of the pollinators. The pesticides need to be controlled because they affect the habitats of insects.

It may affect the food supply, water or even the nesting of the insects. These are among the areas in which the pesticides have affected the lives of the insects. In matters of food supply, some of these pollutants enter into the plant and they will be consumed by the insects thereby entering their bodies. More like humans the chemical once in the body, is able to alter their immunity or even their digestive system thereby harming them. This may affect how they feed as some of them may opt to look to feeding on something else. In the case of the bees, if the food affects their bodies and it may affect the quality of the honey. To some extent it may improve the quality of the honey while to some extent it may affect they lowering the quality of the honey and this may affect the sales and eventually, the honey may lack medicinal value as it used to have.

The same case applies to the water supply. If the water supply has chemical contaminants it may affect the contaminated food. It suppresses the immunity of the insects and the insects will be more vulnerable to illnesses thereby making them weak. This is the same case scenario that applies to insects. Also, the inorganic pesticides have effects on the beneficial insects. They may also need to be regulated when used. In most cases what most farmers use is the IPM on their farms. They apply the insecticides only when there is a threshold economic damage that has been crossed. In the case where there are cover crops that have been planted and are used in rotation with the insecticide-treated cash crop insecticides may be an issue of major concern. The potential risks that have been imposed may be minimized through the implementation of the IPM. It may involve starting the process of treating using the insecticide that have minimum effects in a bid to attain the insect management that is required.

The minimization of the harm imposed on good bugs may be achieved by following the instructions on the labels, hindering the applying of insecticides on the plants that produce flowers, conducting the insecticide application at the morning or evening hours and choosing chemicals whose residuals are at a minimum level. The challenge with the protection of the beneficial insect's conservation is the fact that there is the presence of systemic insecticides having permanent chemical residuals in the soil as well as plant tissues. These residuals are taken up by the plant and are taken up by the plant tissues and travel in the plant vascular system. This leads many components of the plant to be toxic to most insect pests.

There are precautionary steps that are taken up. It involves the insect's habitats not being placed on land that has been contaminated by systemics. This implies that with the lack of scientific proof that remains from initial utilization systemic insecticides do not endanger the insectsthat pollinate flowers. It may be better to make an assumption that there may be a potential danger to the insects. The growers of conventional corn and soybeans may focus instead on the conservation efforts on hedge rose and roadsides that have not been sprayed by the systemic insecticides. They could also make the decision to purchase untreated seeds known to the suppliers. It has been established by the EPA that there are lesser needs of the importance of pre-treating soya beans. In the case where enough growers purchase the untreated seeds they there are chances their availability will increase.

Also, the cover crops as discussed earlier need not be treated using insecticides of any class. There is an exception made in scenarios where there is a cover crop that is utilized for an important use including cattle feed which is should be hindered from serious damages from pest. It is important to know that the treating of cover crops by using pesticides is not common. It is important to sustain cover crops from nearby Insectide drifts. Utilizing of insecticides to stick to the recommendations listed in the label. As there is a need to minimize the use of pesticides, it is important to control the numbers of insects that are present in the terrestrial environment.

6.10 AVOIDING PEST INCREASES

There is research that is required, it has been a strong indication that a high indication that a variety of cover crop cocktails may significantly minimize insects by adding to the population of the important predators as well as parasitotic pests. The given species of cover crops will lead to increasing numbers in unneeded crop pests, through the provision of a certain range of tools rather than plants that may accommodate variety of predators in the population. In a case where there is a need to reduce the potential increase of crop pests, one should be cautious when considering cover crops that have a close relationship with species of cash crops. For instance, in the brassicas including cabbage or broccoli are major cash crop then one will be required to reduce the utilization of cover crops including radish, mustard or turnip. This is because they could house similar diseases and pests in the case of cash crops. In the SARE-funded project, it was noted that the clean stand of phacelia facilitated a home for a tarnishing plant bag.

In the large scale fields crops, cotton is at the top of the list for susceptibility to several major pets, tarnished plant bag, coot on boll warm, cotton aphid, different stink bugs and the tobacco bollworm are amongst major destructors for the cotton growers in southeast. The plan utilized in managing may lead to a dip in the population of the pest with no need for the dependence of insecticide will beneficial. Among the successful strategy was visualized in the SARE- fund research project. This project researches the utilization of cover crops increasing the population of pest's predators who feed on a number of insects. The investigation is made on grounds that most important insects may require alternate sources of food.

The beneficial insects also require vegetation to lay eggs or hibernate over the winter. These areas may also be used as breeding grounds for pests allowing them to thrive in numbers ,therefore, sustaining their numbers in the ecosystem hence there is the striking of a balance. Pesticides are designed to control unwanted plants or animals may not be able to distinguish the beneficial insects and those that are not. Therefore, when applying the insecticides one should be considerate when using them as they may result in unintended consequences and disrupt the natural systems that sustain us.

6.11 CHALLENGES IS MINIMIZATION OF PESTICIDES

Some of the challenges that have been faced with the minimization of the pesticides include the overreliance of people using the chemicals in destroying the insects. The other challenge is that some people get away with breaking some of the laws concerning with the use of pesticides. The other challenge may be the fact that people are not following the instructions that have been indicated on the package thereby their use may be excessive. When these problems are solved then there will be conservation of insects.

In conclusion, the insects can be conserved through the minimization of pesticides. This is majorly through the adoption of other methods that are referred to as alternatives to pesticides. These are a major way in which the use of pesticides can be minimized. Some involve the use of biological methods such as the introduction of predators and also the planting of cash crops. There is also the use of physicals methods such as trapping of the insects and burning of the fields among other physical methods. The use of pesticides is usually the last resort or in the case where the population that is to be controlled is relatively large. Other exceptions are made to allow the use of the chemical method. Some rules have been outlined when using the given pesticides. The most important thing is that the following instructions that have been listed on the packaging of the pesticides. It is also important that the appropriate pesticide is chosen for the insect that is causing the problem so that there is no destruction of the beneficial insecticides.

It is also important that the government and the EPA ensure that they play their role significantly with regards to the authorization of the restricted pesticides. They should also be aware while giving the licenses to the people handling the pesticides. For people who are not complying with the laws concerning the application of pesticides then the appropriate penalties should be imposed on them.

7

Conserving Insects Through Combating Climate Change

CONTENTS

7.1 INTRODUCTION

Every creature on this planet has a purpose to fulfill. In our mission of fulfilling our purpose as creatures, we come across several challenges. Some of the challenges we bring them to ourselves, others are caused by our fellow creatures and some are caused by nature. The problem we cause to us; we do have ways to overcome them since we were their inventors. Challenges caused by our fellow creatures have two sides; that is some we might solve them; others we need the help of our fellows to solve them. Problems caused by nature always have a way of reversing themselves or in most cases, all creatures are forced to find ways to adapt to the changes for no one is able to reverse them, only Mother Nature can have mercy on creatures and solve them for us.

The same does happen to insects. Many individuals consider insects to be of no use to the universe. Without insects, the ecosystems would never be balanced. We need them as much as they need us. It is true most insects cause negative effects on human life with the positive effects being outweighed by the negative ones. This doesn't give us a reason to destroy. Sometimes we are forced to reduce their numbers, so as to reduce the negative effects, which is okay, but we have no right to completely destroy them or their habitats. They do have rights to survive for many generations to come.

As human beings, we might destroy them in different ways. Some of the ways we are always sure are designed to destroy insects. Other ways are not meant to destroy insects, but they do end up destroying insects. Issues like pollution of the environment affect insects and their habitats in diverse ways. Some insects are forced to move to new habitats, others die in their original habitats among other negative effects. One of the major issues that have really destroyed insects in large numbers is climate change.

7.2 CLIMATE CHANGE

The phrase climate change which is also referred to as global warming by many is apparently not a new phrase to most people. It is so enormous that it is said to be the biggest disaster of the entire globe. Not only human beings but also tiny creatures such as insects are facing the wrath of Mother Nature. Human beings are known to be the most evolved among all creatures around the globe. They have quite high intelligence compared to most creatures. A bigger percentage of human beings have used their intelligence to come up with ways to at least perfect their lives. They have gone to the extent of using anything they come across, so long as it improves their lives without having to consider the life of other creatures around the globe. Human beings have owned the globe without the thought of "there are other creatures living alongside them" crossing their brains.

For those whose believe in supernatural beings, more so the Christians, they believe that human beings were given dominion over the earth. But their greed has led them to neglect mother nature. They are torturing their fellow natural creatures, be it animals, insects, plants, name them all. The torture is not really direct, it is indirect torture. The torture is indirect in that, the activities being carried out by human beings are slowly degrading the nature of the globe. Mother nature is quite angry with humans. Despite the fact that humans were given dominion of the earth from the word go, mother nature has awakened to remind humans of their original responsibility of conservation, preservation, among other roles. One of the mechanisms used by mother nature is climate change. She has decided to give a hot blow on our faces. The earth is experiencing quite high temperatures, which are not only affecting insects, but they are also affecting human beings in general.

For those who do not believe in creation theories, climate change, as the name suggests, is the change in the normal climatic conditions of a place which ends up affecting the normal life of a certain region. Most believe climate change only has to do with increased temperatures, but in some areas, it has to do with the drop of temperatures.

The globe is known to have several atmospheric layers. One of the layers, commonly known as the ozone layer, is known to help in the regulation of temperatures received from the sun. It is quite a delicate layer for a minute thickening of the layer can cause diverse effects on the earth's surface. As stated earlier, humans have gone deep into the manufacture of new products. This has led to the establishment of industries all over. During the manufacture of different products, some waste materials that are formed during the manufacturing process are released. The release mechanism is different, however, the main waste release mechanism known to destroy our atmosphere is the release of toxic gases. The toxic gases alter the composition of the atmosphere. Carbon IV oxide, which is one of the major toxic gases released from industries, is known to build up in the atmosphere creating a greenhouse effect. Carbon IV oxide can also be released from the burning of fossil fuels such as coal, oil, charcoal among others. The greenhouse effect is known to harness energy from the sun and generally raise the temperature of the earth hence climate change in most areas around the globe.

7.2.1 Causes of Climate Change

Climate change is not only caused by the industrial revolution around the globe, there are several human activities that have greatly contributed to the change in the climate. Apart from human activities, climate change can also be caused by natural effects on the earth's surface.

Natural causes of Climate Change: Many researchers concentrate on the human cause of climate change, ignoring the natural causes. The planet earth is not a dormant planet, in that it does not change its original state. As its ages with time, from the first day of its existence on the universe, it changes its state either for good reasons or bad reason. This is the same with all creatures, from our first day of existence, we are known to grow. Our growth does have some benefits in that we no longer have to depend on our parents, and we are able to look after ourselves. Our growth does have some drawbacks, during our growth process, we come across challenges that might lead to our extinction, but all in all, we still have to go. So does the planet do, the globe has to change, some of its changes will be of positive impact while others will be of negative impact.

The earth is known to follow a certain path as it revolves around the sun. This path is known as the orbit. During the revolution period, the orbit is known to change its cause once in a while. Scientists have come up with three cycles which support the change of the orbital path of the earth. The cycles have all being categorized under the name Milankovitch theory. One of the cycles changes is that the earth orbit changes its shape affecting the usual exposure of the earth to the sun. The other cycle is that the earth tilts a little bit every once in many years from its normal axis angle to give it a different exposure to the sun. The last cycle is that the earth does wobble on its axis once in a while affecting the angle at which it is exposed to the sun. All three cycles of the Milankovitch theory are known to change the climate of the earth. This is because they take place after thousands of years hence leaving the earth in a different exposure angle to the sun. For example, the northern latitudes can be affected in that the solar radiation is more in winter periods and less in summer periods. This means that the winters will not be normal winters for temperatures will be high causing melting of glacier and the retreat of glaciers and ice sheets and the summers will be quite cold than usual.

Natural volcanic eruptions are known to release different materials. Some discharge carbon IV oxide which is known to cause greenhouse effect hence increase in the earth temperatures for the greenhouse gases are known to affect the composition of the atmosphere. Some eruptions might release dust, volcanic ash, sulfur dioxide and aerosols. Aerosols which are more of a mixture of liquids and solids are known to float in the air hence scatter the normal solar radiation causing a cooling effect during the period they will have to last in the atmosphere. As for sulfur dioxide, it is known to mix with the moisture in the atmosphere hence the formation of sulfuric acid. The sulfuric acid is distributed all over by winds. Sulfuric acid is known to behave more like an aerosol while in the atmosphere hence its blocks enough solar radiation on the earth's surface causing a cooling effect around the globe. That's how volcanic eruptions bring about climate change.

The total amount of solar radiation received from the sun does vary from time to time. The sun is known to have several sunspots. Sunspots are parts of the sun which are quite dark because they have high magnetism effects which prevent the flow of gases that move heat energy from the sun interior hence their temperatures are quite lower compared to the other surfaces of the sun. The total number of sunspots does vary after several years. When the sunspots are high in number, the sun is said to emit more radiation to the earth since the gases that move its heat energy are being altered. In periods of minimal numbers of sunspots, the sun is said to emit less radiation for most of its heat energy is being distributed evenly in its surface with minimal energy being less for radiation purposes. This clearly shows that, naturally, the sun can influence the climate of the globe with the varying number of sunspots after several years.

The earth is made up of several crystal plates. The crust is basically the outer layer of the earth. These plates are the ones that lead to the formation of different continents. This happened in that; the earth was originally a single whole piece that broke into different pieces with time. The different pieces are the continents. The movement of the crustal plates has not ceased. The plates continue to move. This movement leads to the location of some of the plates on different latitudes which affect their normal climatic conditions. This is evident in that some remains of species known to survive in certain latitudes are found in new latitudes which are totally from each other.

Human Causes of Climate Change: Human beings do also contribute to Climate change. Most of the activities conducted by human beings end up affecting the climate of a certain region in one way or the other. There are activities that are known to directly contribute to climate change and there are other activities that end up contributing to the change without the awareness of human beings. All in all, human influence on the climate system is quite clear.

The burning of fossil fuels such as coal, charcoal, oil among others is known to release different gases that contribute to greenhouse effects. Such gases include carbon dioxide, methane among others. The greenhouse gases trap heat energy from the sun increasing the general temperatures of the earth.

Burning of oils is brought about by cars. Human beings manufactured cars to help in their movements from one place to another. Most cars require some oil products either diesel or petrol. The oil undergoes in the engine part of the far so as to produce enough energy for the propulsion of the car. During the combustion of oil, some toxic gases are produced. The toxic gases are later released through the exhaust pipes hence increasing greenhouse gases which cause the greenhouse effect all over the world.

The cutting down of trees increases the concentration of oxygen in the atmosphere. Naturally, trees are known to take in carbon dioxide for the formation of plant food and release oxygen during the day and in the vice versa does happen during the night hours. Lack of enough trees in our surroundings means that carbon dioxide will highly be concentrated in the atmosphere for the trees available would not be able to absorb it. Since with deforestation carbon IV oxide will be in high percentages, it means that it will lead to greenhouse effects where solar energy is harnessed in the gas particles hence increasing the temperature of the earth. Not only does cutting down trees cause increased concentrations of carbon dioxide in the atmosphere, but it does also leads to a change in rainfall patterns. Trees are known to regulate the regional rainfall hence preventing flooding. With fewer trees, the possibility of minimal amounts of rainfall will be possible and if large amounts of rain do occur, flooding strikes the region. With deforestation, the land will be able to absorb more heat energy from the sun. This is because there will be minimal trees to first absorb the heat energy before it gets to the ground level of the earth. Deforestation does also leads to change in that landscape of a place hence the earth will not be able to reflect heat energy from the sun; instead, it will harness it all and lead to an increase in the temperatures in the lower layers of the earth's atmosphere.

Burning of fossil fuels not only causes high temperatures, but it is known to cause cooling effects. For example, if the gases released have a high concentration of sulfur dioxide, the sulfur dioxide is known to mix with the atmospheric moisture and form sulfuric acid. Sulfuric acid behaves like an aerosol while in the atmosphere hence reflecting sun rays back to space. Reflection of sun rays causes a cooling effect in the surrounding.

The manufacturing process of different products in our industries does lead to the production of toxic waste materials. Some of the waste materials are either gaseous, others are in liquid form with the rest being in solid form. Not all industries follow the right procedures of treating such waste before their release into our environments. When it comes to treating gas waste, there is no guarantee that the results will be toxic- free. Such gases once release in the atmosphere they do have their concentration of greenhouse gases which in turn lead to greenhouse effects.

The treatment of both solid and liquid toxic waste is being done through a process known as incineration in most countries. Incineration is the burning of toxic waste materials in a tightly closed container that has high pressure. The waste materials are turned into pure ash which remains in the incinerator. During the combustion process, some toxic gases known to be greenhouse gases are produced. No matter the extent of cleaning, they do end up retaining some percentage of greenhouse house gases which upon release into the atmosphere cause the greenhouse effect of global warming.

The dumping of waste materials in landfills seems to be less dangerous to human beings. What most people neglect is the fact that the continued accumulation of waste materials does lead to combustion. The combustion process of toxic waste materials leads to the release of toxic gases that contribute to greenhouse effects.

All creatures require food for their survival here on earth. This is evident in that, human beings, who are the most evolved species on planet earth, major most of the activities purely to food production. Agriculture processes include the production of food, its storage for future use, its processing to come up with better food products, its packaging, its transportation and its preparation, that is, the cooking of raw food materials. All these processes do contribute to climate change in one way or the other. For example, when it comes to food production, that is the farming activities, large amounts of synthetic fertilizers are used. Synthetic fertilizers are known to be great contributors to the emission of greenhouse gases hence climate change. The large scale farming which is practiced in most countries leads to the clearing of large acres of land, which is more of deforestation hence increased amounts of carbon dioxide in the atmosphere. When it comes to large scale farming, advanced machines are used which contribute to emissions of carbon. The processing of food is done in industries which lead to the release of toxic gases in the atmosphere. Packaging of most processes foods is done using synthetic materials, which after the use of the food products are dumped in landfills. The combustion of synthetic materials is known to cause the emission of greenhouse gases.

Livestock farming has also contributed to the change of climate. Most ruminant animals are known to digest their food using a fermentation process known as enteric fermentation. This fermentation process leads to the release of high volumes of methane gas which is a greenhouse gas.

With the industrial revolution, most individuals are migrating from rural areas to urban areas of their countries in search of greener pastures. Continued migration has led to overpopulation in urban centers. With overpopulation, most hygienic practices are neglected to lead to the accumulation of waste all over the place. As we all, accumulated waste does produce toxic gases during its combustion process. The toxic gases are part of the greenhouse gases which cause global warming.

Examples of Greenhouse Gases: Most people know carbon dioxide and sulfur dioxide to be the only greenhouse gases. However, there are several greenhouse gases.

One of the gases is methane gas. It is known to account for approximately sixteen percent of the emitted greenhouse gases. It is produced mostly in agricultural and petroleum industries. In the agricultural part, it is mostly produced from the digestive systems of grazing animals, manure management industries and rice cultivation areas. Methane gas is also released during the decomposition of waste materials in landfills. Its greenhouse effects are said to be more diverse compared to other greenhouse gases.

The other greenhouse gas is carbon dioxide. It is produced during the burning of fossil fuels like coal, oil and some gases.

Nitrous oxide is a greenhouse gas emitted from commercial fertilizers which are used mostly during farming activities. It is also known to accumulate during the combustion of fossil fuels, in nitric acid production, and the combustion of biomass.

Chlorofluorocarbons are one of the toxic greenhouse gases which are mostly found in our homesteads more so in home appliances such as the refrigerator among others. They are associated with diverse atmospheric effects such as the depletion of the ozone layer.

The last greenhouse gas is the sulfur hexachloride. It is primarily used in the manufacture of dielectric materials such as dielectric liquids which are mostly used in medical procedures. Since they are quite good insulators, they can be used in the manufacture of electric insulators for transformers among other things. They are produced upon combustion of their many products.

7.3 EFFECTS OF CLIMATE CHANGE ON INSECTS

When the discussion of effects of climate change on animals is tabled out, the animal species that come into mind are the polar bears, fish among others. Insects are rarely considered to be affected by climate change at all. However, according to recent research, insects are indeed the most affected species of all the animals. This is because insects are cold- blooded in nature. By being cold- blooded, it means that any slight change in the temperature of their surrounding will completely alter their day to day lives. Insects biological processes will be altered if the insects are no longer in suitable temperatures that support them. Insect reproduction, survival mechanisms, their behavior, their structural features, their interaction with other living organisms be it their fellow insects, prey, predators, name them all will all be affected. The alteration of the normal way of life to insects will in turn affect the other species. Some animals might end up being extinct for they relied on insects for food, pest regulation which is aided by insects will also be almost impossible hence increase in insect -transmitted diseases among other effects. Flower pollination, which is a major activity that relies almost purely on insects will also be altered (Bjorkman & Niemela, 2015).

Due to the continued rise of temperatures, regions located in higher altitudinal levels will be dominated by pests. As we all know, most regions in high altitudes almost all over the world are used to grow crops. With the new domination of those regions by pests, it means crop destruction will increase hence increased numbers in more and more insects.

Changes in climate have influenced the flow of genes from adult insects to their going ones. The periods within which the climatic conditions of a certain region are known to serve as a window for evolution to most insects. This is because insects are forced to evolve so as to fit in their new climatic conditions for their "cold blood" that entirely depend on temperatures for their survival. Due to the evolution process taking place, it means that the phenotypes from adult insects to their young ones will keep on changing until they will be totally different from their earlier generations.

Due to climate change, habitats to most insect species will be altered. Some habitats due to rising temperatures might end up drying up. Some might lack enough supply of food around their habitats. This means that they will be forced to migrate to new locations. When it comes to all aquatic insects, be it marine insects or freshwater insects, a slight increase in water volumes in their aquatic habitats might lead to their destruction hence forceful movements to newer grounds (Bjorkman & Niemela, 2015).

When it comes to insect species interactions, most individuals believe that they are constant in that they have clear boundaries and they are never meant to interact. Some species boundaries might allow interactions while others do not allow interactions. With climate change, insects are forced to interact among their different species. For example, in situations where habitats in an entire surrounding are completely destroyed, different insect species will be desperate to an extent that they do not care about their boundaries forcing them to join forces and locate new habitats which have a quite good supply of food and have suitable temperatures for their survival.

The general behavior of insects will be altered. Insects known to fly might end up opting to crawl for the sky temperatures will have skyrocketed to an extremity such that the insects cannot bear them. Aquatic insects known to spend most of their time submerged in water end up choosing to float on water for the water temperatures are quite high preventing enough circulation of oxygen to support their survival. They float most of the time so as to harness enough oxygen from time to time for their survival in submerged conditions.

Almost all insects are known to undergo a certain life cycle during their development. Some are known to go through the complete metamorphosis as others undergo incomplete metamorphosis. The complete metamorphosis is said to have four different life cycles before the insect is termed as an adult insect while the incomplete metamorphosis is known to have three life cycles for the insect to be said to be an adult insect. Since an insect is cold -blooded, the life cycles, be it the four of them or the three of them are temperature dependent. With variations in climate, it means that the life cycles will be affected. This is because the availability of necessary conditions for the thriving of insects will differ. The availability of light, temperature, moisture and food will be different from the earlier periods. Insects known to withstand extreme changes of the seasons, for they have been in existed for ages, might end up not being able to withstand any more variations hence low metabolic activities. This means that some insects will have to take a short time than normal for them to transition from one phase to another. With increased instars between the different insect life cycles, it means that the generations known to be produced in one reproduction year will be reduced. Sometimes, with climate change, the instars might be reduced, meaning the insect generations in one year will be doubled up (Johnson & Jones, 2017).

In terms of insect body size, climate change has played a major role. Insects known to be originally small in size. This happens in that; the development of insects is affected by latitudes. Insects that move from one latitudinal region to another experience different growth rates. Climate change is known to force insects to move from one region to another. This means that they will have to shift their location hence the change in development rates. The development might be brought about by the availability of quite favorable conditions in one latitudinal region compared to the other leading to increased growth rates and large sizes of the insects. When it comes to small- sized insects, it might mean that they moved from a region that favored their growth to a new region where they are forced to strive for their survival.

An increase in precipitation in a certain region can affect the population of insects. With high precipitation rates, it means that the availability of food for insects will be plenty. The insects will in turn end up being stronger and with minimal striving to do. This will encourage their reproduction activities hence the reproduction of several insect generations per reproduction year that the usual number. More generations will mean overpopulation of insects which will in turn increase the competition for survival resources in their surroundings (Johnson & Jones, 2017).

As the climate of the entire globe continues to change, so does the distribution of insects in it. Aa states earlier, insect species boundaries are being broken. Insects from different species are finding themselves residing with each other in almost the same habitat or having to interact more frequently in their territories. The vice versa is also possible in that insect species are known to interact are ending up being enemies due to competition of resources. In the sharing of habitats, it might lead to the formation of new hybrids. Hybridization occurs in that different insect species happen to exchange their genes hence new insect hybrids are formed. The formation of new hybrids means that new hybrid zones will be formed. The interactions of the new hybrids with their source insects are said to be different in that some insects might interact quite well with their hybrids as others create boundaries with their hybrids. In terms of distributions, insects known to dominate in wetlands seen dominating regions of quite high temperatures. Enmity is being seen between new hybrids and their parent sources due to the never -ending competition for resources such as food for they tend to feed on the same foods, competition for breeding grounds among other resources (Ziska& Dukes, 2014).

Most insects are known to depend on plants for food hence termed as herbivorous insects as other insects being known to turn plants into their hosts. With a change in the climatic conditions of most regions, plants are affected. In times of high precipitation, plants are known to thrive quite well hence increase in insect population and while in periods of high temperatures, most plants end up drying hence minimal supply of food for insects which leads to a decrease in their numbers. The increase in temperatures does end up pressurizing most herbivorous insects to migrate in large masses to new locations hence the exposure of quite healthy plants in different locations to more novel pests. Still, host plant shifts might lead to increased or decreased interactions between different insect species or maybe alter the development rates of insects in general.

Coupling procedures in insects are greatly affected by changes in climatic conditions which does pose a major threat to the reproduction process of insects. Most insect species are known to mate only at certain temperatures. If the temperatures are quite high than usual or quite low than usual, it means that insects will end up not mating. The communication means between insects is said to also depend on climate. For example, the communication between insect mates is said to be thermally sensitive. If the temperatures do not favor communication, it might mean that the period during which insects have the coupling energy might end without any coupling having taken place. Still, under communications, some insects used songs as a sexual signal in long- distance mate location and for courtship periods between insects and the declaration of male to male context to win over some female insects for coupling purposes. However, several insect song characteristics such as the pulse rate, frequency, pulse length, interpulse length among others are all temperature -dependent. With the temperatures varying, it means that these characteristics will not be conveyed appropriately from one insect to another. This is a great disadvantage to most insects for the song characteristics are used to identify appropriate mates and with improper deliverance means, it means that most insects will end up not coupling for lack of proper selection of their mates. The actual coupling process is also temperature- dependent. However, with varying temperatures, it means that the temperatures that favor the coupling process might not be suitable hence most insects might end up not coupling at all, more so the female insects for they are known to be the ones who prefer certain climatic conditions for coupling. Under insect songs, with the variance of temperatures, the wrong signals might be conveyed since most insect songs are quite similar with only a slight characteristic differentiating them. This means that insects will end up coupling with the wrong insect species leading to the formation of new insect hybrids. Unfavorable temperatures, more so for female insects, might also lead to mismating, which is mating of different insect species, hence the production of new insect hybrids (Ziska& Dukes, 2014).

7.4 CONSERVATION OF INSECTS FROM EFFECTS OF CLIMATE CHANGE

With the ever -growing rates of climate change, insect species are slowly declining in number. To most individuals, a decrease in insect population might not seem to an alarming effect. However, with further scrutiny, we are able to see and understand how insects are of great importance in our day to day lives. Without insects in our surroundings, the ecosystem would be imbalanced and this is a great disadvantage to mother nature. This is proven by the fact that everywhere we go, there is the presence of insects meaning that we need them as much as they need us. Insects are known to be part of the very fabric of life. It is high time we human beings understand their importance in our lives and be ready to conserve them and appreciate their priceless value to our daily lives (Collins, 2012).

As stated earlier, insects are all over our surroundings, from freshwater ecosystems to dry land to terrestrial habitats, and some to the marine areas. They actually make up at least seventy percent of the earth's entire population of organisms. They are the many that few of us are able to see. Insects help in pollination processes which lead to the production of the sweetest fruits known to the human race. Frogs, fish species, birds and several small mammals are known to depend on insects for their survival. Basically, insects are the fundamental species of almost all our ecosystems.

The insect structures are said to be quite delicate for each part of its body entirely depends on the conditions of their surroundings. Generally, insects are known to have three body parts, the head, thorax and the abdomen.

Each of the three body parts is composed of several sub-structures for its survival. As for the head, it is packed with sensory apparatus which can only function effectively under suitable climatic conditions. The thorax is fitted with legs and also wings. In quite extreme temperatures, the insect wings are said to be non-functional for how will it manage to sore so high under such extreme temperatures. The legs, if the insect is affected by extremely low temperatures, they might not be able to help the insect in crawling from one region to another. The abdomen is known to specialize in the digestion processes of an insect and also reproduction procedures. As stated earlier, both reproduction and coupling processes are dependent on temperatures. If the temperatures are not suitable, it means that the main roles of the abdomen will not be carried out effectively. As for the mouthparts located on the head part, insects are said to have at most three mouthparts. The mouthparts are used for feeding purposes with each mouthpart being specialized to feed using a certain mechanism. Some are known to suck food, others are known to bite and chew, others are known to burrow among other mechanisms. In times of attack, the mouthparts are known to produce chemicals that will make it hard for their predators to attack. The chemicals can also be used as a means of communication between the many species of insects. However, with the ever -varying climatic conditions, the chemicals might not be quite effective for their composition might be altered by either low temperatures or high temperatures (Letcher, 2009).

When it comes to interactions, insects are known to interact a lot and in large numbers. It is said that there are approximately five million different species of insects. Almost a fifth of them is said to interact every now and then. However, with the variation of our climates being experienced nowadays, their interactions might be altered.

With the ever- increasing human activities which are treating the occurrence of massive ecocide, it is high time we, as human beings act and do our best in reversing the effects of our deeds. Insects are the ones facing a great danger of extinction and we need to improvise ways that will preserve their existence for many generations to come.

Insects have been known to survive all the odds thrown to them before by moving around the different ecosystems. However, with the current climate change being experienced almost all over the globe, it will be hard for insects to move around, they only have a few options, which are not quite viable unless we act. They can choose to either adapt to the changes, which will leave most of them dead, they can move around which is a decision that is quite dangerous for global warming has not only affected one region, it has affected the entire globe and their last option is to give and die. All the options are not that viable as stated earlier, so it is up to us as human beings to act and conserve our insects (Letcher, 2009).

For us to conserve insects, we should be ready to address all issues which have to do with climate change, starting from the destruction of landscapes due to the tree cutting activities done by human beings. Addressing climate change should be a global movement for not only a certain region should be saving their environment.

For insects to be saved from the diverse effects of climate change, we as the human race should be able to address and invest in different fields that will help in combatting climate change effects. One, we need to address the beliefs that human beings have concerning insects and encourage them to engage in more activities that will preserve more and more insects. Two, we should invest in research activities that will help us in preserving insects. Three, we should invest in our ways of interacting with insects daily to ensure that our activities do not expose them to the risk of being extinct species. Four, we should be able to practice different activities of reversing the effects of climate change so as to help save many insect species all over the world. Five, strict policies on insect conservation should be implemented to save insects from climate change which is mostly caused by us, the human race. And last but not least, we should be able to frequently evaluate how our insect conservation techniques through combating climate change are saving insects (Letcher, 2009).

7.5 UNDERSTANDING OF WHY CONSERVE INSECTS

Out first role in insect conservation is the ability to question ourselves on our main motives of preserving the insect life. It is true that over the earth's history, many animal species have become extinct due to natural causes. Insects have to become extinct due to natural causes. Climate change is also caused by natural activities on planet earth, but we should have in mind that the rate at which climate change has been able to spread almost all over the world is because of human activities. So it is our call to not let mother nature punish insects for mistakes we have done. One, we should have in mind that, we are one animal species of its kind, we are known to be the most caring animal species on planet earth. So it is our responsibility to care and conserve our fellow creatures, the Insecta. Two, we might not have been able to value the importance of insects in our lives. It is high time we value and appreciate the priceless worth of insects in our day to day lives. And by doing so, we will have enough motivation to strive and do our best in conserving insects from the effects of climate change. Those two reasons are enough foundation, enough push, enough encouragement, enough motivation, to make insect conservation one of our first priorities as the most evolved and most caring species of them all on the earth's surface. Our third push to conserve insects is that, as the human species, we have already witnessed the diverse effects of the decline of our major ecosystems due to natural causes. It is high time we ask ourselves, what more effects are we going to experience from the negative effects being caused by our daily activities. With this in mind, we will be able to do our best and reduce our daily activities and be able to focus more on reversing the effects of our destructive activities hence preserving insects. Our most important push is being able to perceive insects as our fellow buddies in the strive for a better life on this planet.

7.6 RESEARCH MISSIONS ON WAYS OF CONSERVING INSECTS

After having enough push to conserve as to why we should conserve our insects, our second role us conducting research. The research will help us in obtaining more information about insects on how they are being affected by climate change. For us to understand the challenges insects are facing due to climate change, we should have adequate information about their day to day ways of life, how different species behave and their interactions and finally, the approximate number of insects for a better understanding as to how we are going to manage them all. For us to be able to conserve them, we should also have adequate information about the environment in which they survive. This will be a bigger task for we are supposed to research on the perfect climate conditions for each insect species, the food they feed on, their life cycles, both their respiration and reproduction mechanisms, their feeding mechanisms, their prey, habitats in which they perfectly survive in among other several issues to do with insect livelihood. When it comes to the insect's life cycle, we should be able to understand what the perfect conditions are for a certain insect phase to be attained and how the different instars of insect life cycle thrive (Collins, 2012).

Insects are known to be able to thrive in more than one habitat. In our research activities, we have the task of gaining information on the different habitats a certain insect species survive in. For a better understanding, we should be able to gather information on the differences and similarities between the several habitats. With all this information we will be able to understand how different habitats for different insect species should be preserved.

The movement of insects is known to be frequent in that, they are known to always be on the move now and then. Insects move in search of food, in search of mates, in search of breeding grounds, in search of new habitats due to overpopulation issues among other needs. During their movement, there will come a time whereby they will have to make a stop. Due to our activities, most stops being made by insects have occurred on grounds that have been known to end up destroying them instead of preserving them. It is our role to research the perfect conditions which favor the survival of any insects. With this information, we will be able to leave several areas around our surroundings that have the perfect conditions for the survival of insects, be it in the rural areas or urban areas. These areas should be perfect in that, in case the insects happen to breed in them, the chances of the production of viable offsprings are quite high hence preserving many insect generations for ages to come (Collins, 2012).

Still, under research, we need to understand the different species of insects in terms of their reproduction mechanisms. The survival of insects in certain conditions does vary from one species to another. There are some species that are known to exchange the alleles and come up with new hybrids. Most of the new hybrids are said to be well adapted to the new effects of climate change. However, the flow of genes from one insect species to another is known to be affected by climate change. This gives us the role of gathering information on the most effective ways which will support a good flow of genes between different insect species for the formation of new hybrids that will be able to defy almost all the challenges insects face due to climate change. Another research that should be conducted is that of finding ways of ensuring different interactive insect species are not separated for together they will be able to overcome basically any challenge.

We are responsible for conserving the various landscapes on the earth's surface. It has been known that through our activities such as the cutting down of trees might lead to erosion which leaves most landscape in pretty bad shape. For example, marine insects might end up dying due to the accumulation of dust from the collapsing of lands that were once held together by vegetation. Their respiratory systems are blocked by the dust and they face challenges in harnessing enough oxygen hence their death.

It is also clear that the attempts to only preserve an individual species of insects have ended up bearing minimal fruits. This is because insects are known to move from one location to another. It is our responsibility to research the perfect landscape conditions which will support the survival of basically all insect species. Governments should invest in creating a resilient landscape and marine environments that are not prone to destruction by either natural or human causes. We should value our environment for with that it will be quite easy for us to maintain the landscape in our surroundings. With detailed research on the landscape, we will be promoting the growth of insects and conserving them. Preserved landscapes will encourage the interaction of insects hence the production of more resilient insect hybrids to climate change. Not only insect species, but also all animal species on the earth's surface will also be able to benefit from such preservation of landscapes (Samways, 2019).

7.7 GUIDELINES FOR INSECT CONSERVATION

After thorough research, it is up to us to come up with rules and regulations that will ensure that insects are well preserved. For us to come up with policies that will better implement the information from our research, we should be able to view insects from two different perspectives. One of the perspectives is that we should be able to acknowledge insects as part of the biodiversity in which they play a bigger role in balancing our ecosystems. The second perspective is that we should consider insects to be part of our daily lives and value and appreciate the important roles they play in our day to day lives. With this in mind, different policies can be implemented. The execution of such policies should be done globally with the natural level being the main background for ensuring that the rules are strictly adhered to for us to reach our global goal of conserving insects and preserving more and more generations of insects.

When it comes to policies to do with biodiversity, illegal clearing of vegetation and destruction of the environment should be prohibited. This is because insects are part of a large ecosystem. In this ecosystem, we do have other species that interact with insects, be it plants, be it fungi products, be it other animals, among others. If we destroy the environment, all these organisms will all be at risk. With a preserved ecosystem, it is a sure bet that insects will be maintained in good conditions (Didham et al, 2020).

When it comes to disposing of waste materials, a policy that wastes should be treated before being disposed of should be implemented. More so for gaseous wastes. This is because, as stated earlier, gaseous waste is known to contain toxic gases that promote greenhouse effects, which increases the temperatures of the earth. With increased temperatures, most insect activities are altered for insects are known to be cold -blooded hence most of their day to day lives are temperature- dependent. The release of untreated liquid wastes materials in water bodies should also be prohibited. As we all know, there exist aquatic insects both of which, live either in freshwater or marine environments. Untreated liquid waste is known to contain toxic chemicals that alter the composition of water hence changing the water temperatures, contaminating aquatic vegetation and finally destroying the life of aquatic insects.

A policy should be implemented to identifying the most endangered insect species and coming up with a list. This policy should further outline that such species should be collected and preserved in natural habitats which are regularly examined to ensure that they do not end up becoming extinct. Any attempt by any person to destroy such habitats: the culprit should face serious consequences for exposing the endangered species to the chances of extinction. The red list should not be considered as the first priority in insect conservation, for there are other insects still under threat. However, it should be used as a way of noting the progress which has been made so far in preserving insects. The main challenge facing the red list is that only a few insect species have been added to the list conveying the message that we are quite far from hitting our target of saving as many insect species as possible (Samways, 2019).

The other policy that has been implemented is the encouragement of more individuals to join the community of insect specialists. With many educated insect specialists, it is believed that we will be able to save as many insects as possible. The specialists will encourage the growth of more and more insect species and even help in coming up with manually fertilized insect hybrids which will be able to withstand most of the challenges which have been caused by climate change. With enough insect specialists, we will be able to better understand the main threats for insects. The threats such as the loss of natural habitats due to illegal logging, loss of habitats due to increased agricultural activities which have led to the clearing of many acres of land, increased pollution rates, increasing mining activities, among other threats will be addressed in the most effective way hence saving our insects. Understanding such threats will help in planning for better strategies to be employed in the future to prevent any more destruction of habitats among other human activities that expose insects to the risk of becoming extinct.

A policy has been implemented to ensure that mining activities have minimal impact on insects. Technologies are being invented to promote the prevention of distraction of the insects' habitats. This is being done in such a way that, the insect habitat will be protected as the machines navigate for the minerals.

Away from biodiversity, insects are known to be of great importance to the day to day livelihood of human beings. They help in breaking down both living and dead plant materials hence, promoting the fertility of our soils. Most communities value insects for cultural purposes henc,e acting more as the identity among different cultures which we have in our society. Insects can be used by scientists in genetic research. With all this we able to see the value of insects and be able to come up with policies to conserve them.

One of the policies which have been implemented based on insect value is that, when conducting genetic researches, insects should be collected most effectively and conservatively from their habitats exposing less risk to their survival. During the study, they should be handled carefully and returned to their habitats in their original conditions (Kumar, Yogi, &Jagdish, 2013).

Another policy is that communities that feed on insects should be able to rear them in their homestead. This is to ensure that they do not deplete the class Insecta species. They should preserve them in the most effective way to increase their population such that we will have enough insects for feeding on and also for preserving more and more generations of their kind.

Another policy is that farmers should be encouraged to apply minimal amounts of fertilizers in their soils. This is because some insects are known to help in controlling pests by feeding on their fellow insects. They also help in the replenishing of nutrients in the soil by breaking down dead plants that boost soil fertility. Fertilizers are known to have a large percentage of chemical contents which might end up affecting insects known to have their habitats in soil. The chemicals might also affect the normal day- to -day activities of insects. With these negative effects being caused by chemicals, insects known to help boost soil fertility are rendered useless.

Some insects are also known to be of medicinal value to human beings. A policy should be implemented to ensure that they are reared in large numbers in manmade habitats to ensure that they do not end up being depleted due to capture by human beings for medicinal purposes (Kumar, Yogi, &Jagdish, 2013).

7.8 HUMAN INTERACTION WITH INSECTS

For effective conservation of insects, we should be able to understand our interactions with them. This is because in most of the instances in which we interact with insects we leave most of them dead or exposed challenge to them in our day to day activities. However, we should be able to discipline ourselves and ensure that most of our activities are directed towards the conservation of the environment in general. Before carrying out a certain activity, we should be able to question our motives and evaluate if the activity will affect the environment positively or negatively. We should also understand that, insects play a major role in our lives, hence we should value them and appreciate them for they are quite vital in our day to day lives.

As the technological age continues to advance, we as human beings should have in mind the nature of insect life before the invention of any technological equipment. This will ensure that the end products we come up with do not end up destroying insects. For example, instead of having to throw away our old technological devices such as laptops, laptop batteries, old computers among others in landfills, we should dispose them in a more appropriate way to prevent the damage of insect life for most insects are known to reside in the soil.

Most people are known to hold only a few insect species dear to them. However, it is our responsibility that we value each and every insect we interact with more so the ones that benefit us the most. We should value and preserve insects such as bees, butterflies for their roles of beautifying our environment, grasshoppers, land birds, dragonfly among others. As for the bees, we can go to the extent of ensuring that most of our environment is surrounded by flowering plants which will support the bee making prowess by bees. We should also construct beehives and place them in quite surroundings so that the bees have ample time while making honey.

As for the dangerous insects such as the likes of mosquitos which cause diseases to us such as malaria, the narrow bee fly among others which are a nuisance to us, we should not use eradication mechanisms that will lead to their distinction. Technologies should be implemented that will instead of destroying them, we only render them harmless to us but leave the ability of their survival in them. For example, a mechanism has been implemented that mosquitoes will still be able to feast on human blood but their bites will not leave any pathogen on the human body that might lead to malaria. This has been done in such a way that genetically engineered mosquitoes are released to breed with the exotic mosquitoes hence their off springs will be pathogen- free bringing an end to malaria hence, no more use of insecticides to destroy mosquitoes.

As we use insecticides, we should the thought that they are not affecting insects, but they are also affecting us as human beings. This is because, before the insect dies, it will also attach itself to the food materials in our households. Upon consumption of these foods, we will also end up suffering some illnesses due to the ingestion of the poisonous chemicals into our bodies. Also, the destruction of insects by applying large amounts of insecticides we will be rendering them useless to the surroundings more so in terms of pollination meaning that many plants will be left not pollinated hence no produce in our farms.

After gathering enough information through research activities, and implementing policies that help in the conservation of insects, it is high time we put into practice all the implemented policies on insects' conservation from the negative effects caused by climate change.

One of the ways to ensure that the strategies on insect conservation are implemented is by educating the public. Most of the research and study of insects is conducted by insect specialists. And as stated earlier, not all people see the worth of insects, so it high time the public be educated on the importance of insects. The public should be made aware of the benefits we benefit from insects on our day to lives. In terms of the dangerous insects, the public should be sensitized on the proper ways of handling the many harmful insects in our surroundings.

Governments should be able to invest in the manual rearing of insects. This is to ensure that communities that rely on insects as a source of food have the chance to have an adequate supply of insects and also have some remaining insects to ensure that the insect species are preserved for many generations to come. The rearing of insects will also be important in terms of gathering enough insects to be used for medicinal purposes and that some of the insects remain for preservation purposes.

National parks and national reserves should be preserved in the most effective ways. This is because these parks serve as habitats to many insects. The taking over of these lands in these areas by human beings should be prohibited to ensure that the natural habitats are preserved and the insects have higher chances of survival despite climate changes.

After understanding why we need to preserve insects, conducting thorough research, implementing different policies, understanding our different relationships with insects and putting into practice the different strategies of protecting insects, it is high time we have records that will track our progress in conserving insects. Research records should be well documented to track how far the insect specialists have done their research. The implemented policies should be documented if possible in both soft copy and hard copy to ensure that in case any references are to be done, anyone can access them with immediate effect all over the world. The documented policies will also help in tracking how far the insects have been preserved.

8

Conserving Insect Through Removal of Invasive Alien Plants, Mixing of Crops and Insect-Friendly Plants

CONTENTS

8.1 INTRODUCTION

Invasive alien species are non-ecosystem native plants that may cause environmental damage or harmfully affect insect health. In essence, they adversely affect biodiversity's, such as through competition with native plant species, the spread of pathogens, and destruction of indigenous ecosystems and environmental functions that are necessary for insect survival (New, 2016).

Figure 8.1: Invasive plant species devastate the habitats of native insects and prevent them from multiplying.

Source: https://www.cabi.org/what-we-do/invasive-species/

Invasive plants introduced and / or dispersed outside usual environments have impacted on the native ecosystem of almost every terrain on the planet and are among the biggest risks to biodiversity. As of the seventeenth century, invasive plant species have accounted for almost 40 percent of all fauna disappearance, including known causes of insect extinction. Climate change, loss of habitat, human disturbance and, pollution have exacerbated this damage.

Restoration programs to eliminate invasive plants may have a great influence on local plant species. Nonetheless, a new research in the bulletin of the Invasive Plant Sciences and Control shows that recovery has other benefits. Elimination of invasive vegetation growing beside streams and rivers can also advance the biodiversity of water insects, like mayflies, dragonflies, stoneflies and caddisflies

The scientists led a study to examine whether removing invasive Amur honeysuckle undergrowth that grows with upstream water sources in Ohio forests could affect the variety of insects living in the river.

A 160-meter long honeysuckle plant invasion was eliminated, which significantly enhanced the light utilization rate along the river and altered the organic material quality. Researchers examined the flow every month for roughly three years. They found that eliminating honeysuckle had a significant positive effect on the concentration, abundance and diversity of insects living in the river. Obviously, even if simply a small part of the river is restored, the removal of invasive plants will have a top-down effect on the nearby water-insect ecosystem (New, 2016).

8.2 IMPACTS OF PLANT INVASION ON INSECT DIVERSITY

Invasive alien plants have a disturbing effect on local insect biota, resulting in the reduction and even extinction of native species and adversely upsetting ecosystems. They are plants or fungi that enter from outside their natural habitat and establish themselves in the new environment. They breed quickly and compete with local vegetation for nutrition, water and space, plus are among the leading causes of the loss of insect biodiversity worldwide. Species are usually intentionally introduced through gardening and biological control, or involuntarily introduced through terrestrial and water transference, travel, and study research (New, 2016).

With increased transportation of goods and movement, the global economy has encouraged the relocation of plant species across vast expanses and past their natural boundaries. Although merely a small fraction of transmitted plants are intrusive, they have a significant effect on the condition of local plants, insects and organisms that endanger lives and affect food safety and health of the ecosystems they live in.

Eradication is the desired solution once established but it may be quite costly to do. The best solution to control invasive plants is always prevention. Climate change, habitat loss and pollution will exacerbate the negative impacts of invasive plant types on insect biodiversity. Of special concern are remote habitats, such as islands. Invasive plant species are regarded to be the main source of pressure on the global status of insect biodiversity. In particular, alien plants have dispersed rapidly and widely in many parts of the globe, affecting the diversity of local species and ecosystem processes (New, 2016).

8.2.1 Case Study

In South Africa, exotic plants invaded almost 2 million hectares of land, which had a well-known effect on hydrology, mineral cycling and fire cycles. It is estimated that the coverage of exotic plants comprises 27 species, which does not include drought and transformed lands, excluding the Prosopis tree from the arid northwestern region of the country. Exotic acacia tree species make up most of the invaded area, including eucalyptus and pine, cactus and Chromolaenaodorata bushes.

These invasions spread throughout the country, and are more concentrated in the southwest and south, especially in the eastern coastal zone and adjacent inland areas. Overall, there's practical knowledge about the occurrence of exotic plants in South Africa, particularly at a rough spatial resolution. Although their impact on local plant diversity has been well evaluated, few studies have concentrated on the effects of exotic plants on local insect communities.

In this study, it was established that the bearing and extent of the impact of foreign plant invasions on insect communities depended on a range of factors, comprising the scale and density of plant invasions, the phase of the invasion, and the local and biological classification. Invasion of foreign plants into nature or formerly uninvaded habitats involves many major changes to the habitats, which usually have a negative impact on the resident fauna, sometimes even counterintuitively or inconspicuously. Exotic plants may directly change the arrangement and intricacy of the physical environs, thus limiting the opportunities for insects to adjust temperature or hydro-regulate in their micro-environment, or setting obstacles to basic functions like movement, nesting, or finding shelter. Exotic plants can also openly or indirectly influence the food resources of insect communities.

For instance, changes in plant arrangement will directly impact herbivores by decreasing the quantity or host plant quality, while changes in habitat arrangement, microenvironment, and litter or earth properties will indirectly impact the availability of insects or the abundance of their predators. Although the negative effects of invasive plants are generally reported in the literature, there are also reports of positive effects, for instance, by increasing the appropriate habitat or overall resources for the recipient bugs.

More notably, understanding the direct and eventual causes of insect species decline may improve the capacity to predict the response of bug communities in newly occupied regions in South Africa, or when they face accompanying pressures like global warming or habitat changes. These are the basic components of research frameworks on invasive vegetation. These factors will enable plant intrusion science to effectively predict the influence of new and emerging intrusions.

8.2.2 Stages of Invasive Plant Species Removal

Stage 1: Participatory design and assessment of invasive species and ecological system scenarios

The key goal of this phase is developing prospect interventions for the management of invasive plant species, through the practice of participatory scenario planning or (PSP) with other researchers and stakeholders. This is a progressively widely used tool that helps to explore the future development of complex systems, thus providing details for decision-making.

Researchers will choose several invasive plant species of interest, which shall be the center of research. They can choose any invasive plant species between 10 and 20 in number, covering various environments (freshwater, land), invasion stage (from scratch to widespread) and impact (from smallest to huge, positive or negative).

The choice of species will meet the research requirements depending on goals, data and resource accessibility. These studies may center on a subgroup of the available species list. Priority is provided to species already listed or anticipated to be added in the EU's IAS list, which is a catalog of 49 (plus 18 potential candidates) animals and plants that the EU has prioritized management according to Regulation 1143/2014.

Scientists have envisioned two extreme possibilities, that is, invasive plants will affect insect populations. First of all, in one case, exotic species aren't managed, then by 2050, they will be able to occupy the range of potential climates simulated by increasing globalization. Secondly, in another case, globally coordinated activities prevent the entry of new plant invaders by means of border control as well as trade regulations and manage existent invasive plant species through eradication systems and ecological renewal.

Between these two, various situations can be drawn for preserving the number of insects, preferably through a participatory procedure, so that the participants have practical experience in ecosystem management and restoration. For all these situations, the intervention plan will regard all levels of the invasion procedure and the affordability, practicality and adequacy of any substitute management plan.

Stage 2: Assess the positive/negative effects of invasive plant species on biodiversity as well as ecological services.

The key goal of this level is to consider the harmful and beneficial effects of invasive plant species on insect biodiversity and ecological services and to adjust and verify the impact assessment plan. The impact appraisal protocol can be applied to measure the impact of the biodiversity and ecosystem solutions of the invasive plant types of interest from five levels of concern which are Minimum Concern to Large Scale.

The goal is measuring the possible effect on a five-point scale through quantifying the impact of invasive plant species on how many native insect species and uncertainty estimates at a triple point scale of low, medium and high. This consistency can help to compare the harmful and positive effects of an exotic plant to insect populations.

Stage 3: Modelling and charting the impacts of invasive plant species on insect biodiversity and ecosystem services

Within the context of this model, researchers can shape and map out the possible impacts of invasive plant species on biodiversity or ecosystem services, within the present and future climate change situations heading into 2050. Distribution modeling practices can be applied to compare the occurrence of invasive biota of interest, making use of the environmental circumstances of their local and introduced ranges. The models can be applied to envisage the development of species under present and future 2050 situations and identify the areas of greatest concern which are vulnerable to invasion.

The potential impact of invasive species can be calculated by combining species-specific scores with predicted distribution maps and ecosystem service supply plans. First, researchers will collect existing maps from the literature that reflect ecosystem services (example, pollination and water purification, all of which are available on a regional scale), and target insects that will be directly affected by the invasive plant species.

Scientists will then combine species distribution maps that affect specific insect ecosystem services to identify regions with high supply and greater risk of invasion. Since the effect of invasive plant species depends to a great extent on specific circumstances, they must assume that the possible impact of the particular invaders throughout the area is similar, which is a common precautionary standard in invasive biology.

The plan provides a structured assessment of management possibilities for present and imminent invasive species, using semi-quantitative reaction and uncertainty scoring to measure the efficacy, practicality, rate, impact, suitability, and opportunity of the feasibility criteria and possibility of invasion again.

Finally, the potential costs and advantages of alternative interventions in insect biodiversity and ecological services can be compared. The generated data, maps and prototypes will clearly assess the threats to insect biodiversity and ecological services in particular areas, as well as the potential response measures under a variety of climatic and managerial scenarios.

Stage 4: Appraise and map the impact of invasive plant species on insect biodiversity and ecological services.

At this stage, the objective is to identify invasive plants which have not been reported in some areas of the infested region but are expected to change in scope as climate change expands. This can help predict the threat to insect populations or ecological services of invasive plant species under other interventions.

The researchers will first build a database that synthesizes all the ecological effects reported in the research literature, apart from using established protocols to assess the impacts. They will then use the existing occurrence data to simulate the possible distribution of range-changing invasive plants. This approach focuses on emerging threats, and established species that are likely to be invaded under climate change variations (known as sleeper species) are out of the range of this approach.

Stage 5: Impact of invasive plant species on regional-scale biodiversity and ecological services

This level aims to improve empirical understanding of the impact of invasive species on insect biodiversity and ecological services, and their environmental dependence. At the local level, many invaded and non-invaded locations in freshwater, land and marine habitats can be chosen to measure the effect on biodiversity as well as ecosystem services.

Furthermore, after recovery, physical removal experiments that cause minimal disruption to non-target species can be applied to track changes in ecosystem characteristics. The goal is to assess insect biodiversity and ecological services under this general assumption: the location of restoration will be somewhere between invasive and non-invasive treatment. This basic hypothesis is rarely tried in the field; however, it plays a significant role in distinguishing the main impacts of the insect population management program.

8.3 BIOLOGICAL CONTROL METHODS

Figure 8.2: Ladybirds are commonly used by farmers to control the population of aphids.

Source: https://byjus.com/biology/bio-control-agents/

One technique for controlling invasive plant species from damaging insect habitats is biological control, that is, "using their natural enemies" to deal with them. These "biological control agents" may be bacteria, fungus, viruses and parasitic or predatory fauna like insects. Biological management has been successfully used to supplement and maintain the physical and chemical elimination of invasive alien plants, particularly in wood invasive species. If biological control methods are shared with physical or chemical control procedures of woody species, general savings of between 20 percent and 30 percent can be achieved (Reddy, 2017).

When there's a conflict of interest in controlling or eradicating invasive species, biological control can be a great solution. For instance, some acacia invaders have may have ecological value to some insects. Biological control agents can target particular plant organs, for instance, flowers and seeds.

Therefore, whenever an invasive plant has some biological value to certain insects, the agent selected may be an agent that only invades the vegetative regions of the plant or an agent that reduces seed yield, instead of an agent that may damage beneficial sides of the plant.

The seed reducer agent diminishes the reproductive prospective of the plant, inhibiting the spread of the invasive plant, and reducing the follow-up task required after cleaning, whereas still using the plant. In case seeds are required to replant the plantation, it is possible to biologically control the seed orchard in a similar way as protecting other crops from pests to achieve special protection. Even though biological control is a safe and affordable control method, it does require some research and high levels of education and professional knowledge among researchers.

The biological control agent is only released after it's been demonstrated that it has sufficient host specificity for local circumstances, otherwise, it might also cause insect death. Therefore, tested and approved biological control agents do not posture a threat to native insect communities or insect communities in nearby areas. To date, there has not been a situation where the invasive plant biological control agent changes its host plant affinity after release in a new environment to include plants apart from those identified acceptable hosts (Reddy, 2017).

Since 1913, well over 82 biological control agents have been used in South Africa to manage 47 kinds of weeds negatively affecting insect habitats. From these introduced plant species, unpredictable host transfers have not occurred. The reason for biological control must never be to entirely eradicate the host plant population. Biological control agents lessen population density and diminish the vitality and/or reproductive prospects of individual plants.

Appropriate biological control reserves must be retained to maintain a small amount of biological control agent in the cleared area to ensure that the biological control agent and native insects will not die from lack of food. Persistent small amounts of biological control agents will be spread on any regrowth or emerging seedlings of invasive weeds.

Therefore, biological control may be considered as a viable control measure. The benefit of biological control is that it's environmentally sustainable, does not cause pollution, and only affects target (invasive) plants. It is permanent or self-sustaining. If managed properly, it will be very cost-effective. It will not disrupt the soil, nor will it create large vacant places where other exotic invaders can build, because it will not immediately kill all target plants. It can make the natural vegetation in the area gradually recover in the sanctuary of disappearing weeds, which can support more insects (Shields, 2017).

The combination of these factors makes investing in biological control worthwhile. In addition, the research experience and expertise are used for projects in various parts of the infested area. This increases the opportunity for effective control of the entire continent. History furthermore provides researchers with sufficient evidence to successfully apply and utilize particular biological control methods.

8.3.1 Case Study: The Prickly Pear

Opuntia or prickly pear is a successful biological control story. The plant got introduced into Australia during the late 1770s, growing in some regions of New South Wales and Queensland. It did not begin to invade until it quickly spread after the 1893 flood. Biological control began during the early 1900s, where the cactus fruit moth Cactoblastiscactorum was planted in 1926 taken from the pear's original habitat in America. To date, the moth is still controlling prickly pear numbers in Australia, ensuring that the exotic moth species not only thrive but also allows for other local insects to do the same.

Since then, more biological control agents have been used to manage invasive plants. They include Mimosa at the top, Bridal Creeper in South Australia, Wormwood in Tasmania and parthenium in Queensland. A number of affordable analyses in 2006 showed that for each dollar invested in the biological control of invasive plant species, insect farming industries like beekeeping and society would benefit. This is because of increased production, cost savings of billions of dollars and advantages to human health.

The original costs of biological control programs are usually high. This is because users must find appropriate candidate agents in other lands, conduct quarantine security tests on them, and conform to the regulations regarding the release. However once the biological control agent is released and affects the entire range of invasive species, the cost of subsequent control will be greatly reduced (Shields, 2017).

Fungicide: Fungicidal herbicides are mixtures of fungal spores within a carrier, and may be used on weeds in the same means as chemical herbicides. Spores germinate on plants, infiltrate plant tissue and cause infection, and may ultimately kill invading plants.

Mycoherbicides fungicides, for instance, are inherent in some countries/regions and hence already exist naturally in the environment. In natural conditions, they will not cause sufficient damage to the invasive weeds and thus won't cause damaging effects. Therefore, they are mass-produced and spread using herbicide dispersal equipment in a traversal inoculation manner, resulting in the epidemic of disease and the death of invasive plant populations.

They cannot sustain the epidemic within natural conditions and therefore require reapplication. Because herbicide resistance is native, but only causes low amounts of illness under natural circumstances, they don't pose any threat to non-target native plants and bug populations that depend on them.

The development of herbicide from natural pathogens has made great progress in science, the herbicide can be used to control hakea seedlings as well as mature plants (Hakatak). Many fungi may be developed as anti-mildew herbicides against water hyacinth or mesquite, which have been known to devastate aquatic insect reproduction cycles.

Besides some research has been conducted in the past against this target. In South Africa, the first innovation was the creation of a non-pathogen fungal inoculant for preventing the regeneration of black and tan thorn residue (residues) after felling.

Chemical control methods: Chemical control includes using registered herbicides for terminating target weeds. Most studies on chemical regulation of plant invaders aim to develop safe, discerning and inexpensive herbicide treatment methods that will offer practical control under a wide variety of environmental situations and are attuned with biological management. Herbicides are one of the most practical and resource-efficient apparatuses for treating invasive species. Most common invasive plants can only be treated with two herbicides which are; glyphosate and triclosan.

Glyphosate is largely non-selective, meaning that it kills all objects in contact, including foliage that is beneficial to insects. Triclopyr is meanwhile selective and will not harm monocotyledons (grass, orchid, lily, and so on.). They are fairly benign herbicides, however, improper use can still result in temporary and longstanding health and ecological problems.

When in wetland areas, a special permit may be required when spreading these chemicals on terrain that you terrain don't possess. To discover more about pesticide regulations within your area, you can consult with the pesticide regulation department, which is typically found within the Ministry of Agriculture.

Foliar applications: When the scale of the plant invasion problem is small, a backpack sprayer and even a tiny handheld spray bottle are typically used for such treatment. This is an ideal method for processing large single herb plants, and can also be used for on-demand treatment of specific plants that are hard to eradicate mechanically, like gout grass, purple loosestrife, and swallowwort which often strain the insect's habitat.

It's furthermore an effective treatment process for certain woody species (like barberry, multi-flower rose and honeysuckle). These species grow densely or abundantly in many acres. The herbicide solution must have not more than 5% of the active component, however, it's essential to follow the instructions mentioned on the product sticker.

When alien plants are quickly multiplying, it's best to apply treatment when they bloom or begin to set in real- time, this treatment technique is the most practical. It's been revealed that when existing stems are diced and treated for regrowth, plants are typically more prone to this treatment. This is particularly accurate for Japanese knotweed. Within the next 1 to 2 days when there's no rain, the herbicide can be used to methodically wet the target plants.

Cut stem treatments: There are many unique kinds of stem cutting treatments, but this review will only focus on the most popular. Compared to the foliar application, all such treatments require greater concentrations of active components. A solution of 25% to 35% of active ingredients must be applied for stem cutting, however, it's important to read and follow the label instructions accordingly. In many cases, the proper herbicide glyphosate must be applied alongside Oriental Bittersweet. The treatment can be applied to all alien woody stems, reeds and knotweed that affect insect biospheres. For woody stems, the application is most effective if applied during late summer or autumn seasons which begin from late August up to late November.

Stems must be sliced near the ground, however not too close to avoid losing them. Spread over the herbicide directly onto the cut surface immediately after cutting. The delayed application will decrease the efficacy of treatment. As for small insect habitats, the herbicide may be used with a sponge, spray bottle or paintbrush. For reed and knotweed, the treatment is similar, however, the timing , as well as equipment, used are different.

Mechanical control methods: This requires physical action to destroy or remove invasive plants. Different skills can be used, for example, uprooting, clear-cutting, mowing, barking or peeling. The control approach is only really practicable in the case of sparse infestation or small scale, besides it's indeed feasible to control species that cannot meet the biodiversity demand after cutting. Species that seem to be strong require the treatment of cut stumps or strong seedlings sprayed with herbicides after mechanical treatment. Mechanical invasive plant control can be labor- intensive, and thus expensive, and might cause intense soil disruption or erosion.

When evaluating invasive plant removal programs, mechanical treatment is typically the first consideration. These processes don't need special permits or the introduction of chemicals into the ecosystem, which may negatively affect insect populations. In some cases, such as wetlands, they do need a permit. Mechanical destruction requires a lot of labor and will cause a lot of on-site interference. If not handled properly, it may cause quick reinvasion.

Pulling and digging: If the number is limited, many herbs and some woody plants (up to an inch in diameter) may be drawn or dugout. It is essential to erase as many root structures as practical; even a small part may restart infestation again. Pull the plant through hand or with a digging fork, because the shovel will cut off a part of the root structure and make the plant grow again. To eliminate larger wooden stems (about three inches in width), it's possible to use a wrench.

The tools are available from various manufacturers and are developed to eradicate the aerial parts of the plant and the whole root system. It's simplest to undertake this kind of control during spring or early summer, when soils are humid and invasive plants emerge more easily.

Suffocation: Try to suffocate small seedlings and herbs of the invasive plant. Put two or three layers of UV-stabilized plastic cloth (transparent or) on top of the alien plants, and fix the plastic with wooden posts or heavy objects. Ensure that the plastic spreads at least 5 feet from the invasion edges on all sides.

Next, allow the plastic to stay for about 1 to 2 weeks. This technique can exterminate everything under the plastic, including non-invasive plants. Following removal of the plastic, sow crops such as yearly rye to inhibit new alien plant invasions.

Cutting or mowing: This method is best applicable for places you tour and treat frequently. To be operational, you will have to trim or slice the infested area 3 to 4 times per year for around five years. The aim is to interrupt the plant's photosynthetic capacity by eliminating as much leafy matter as conceivable. Expunge the plants found on the ground then eliminate all debris produced on site. Through this treatment, the infestation may really get bad at first, therefore you should be as enduring as the alien plants because ultimately they will be eliminated.

Every time a plant's part is cut, its root system will become larger, but it must depend on its energy resources to promote new growth. Ultimately, these reserves will become scarce and the plant shall die. This might take several years, therefore once you start, you should continue working on this process. Otherwise, the treatment may be counterproductive and make the situation worse.

8.4 MIXING OF CROPS AND INSECT-FRIENDLY PLANTS

Figure 8.3: Mixed cropping allows for different insect species to co-exist in the same ecosystem.

Source: https://www.gardeningknowhow.com/edible/vegetables/vgen/edible-landscaping-mixing-vegetables-and-herbs-with-flowers.htm

On its own defense, farmland with mixed crops tends to attract fewer pests than farmland with only one crop. Although researchers and farmers have noticed this disparity for some years, still little is known concerning the reasons around it. Some of the reasons for this phenomenon are that insects normally have some perfect nutritional level they prefer, or when plant nutrition is too high or very low, then their performance can be affected. The concern with monoculture is that when an insect prefers a particular crop, then the insect has plenty of food to eat from one place (Konvalina, 2016).

On the contrary, a field with different plants cannot provide a lot of food for insects, therefore it won't get the nutrients needed for survival and reproduction. Single planting resembles a buffet for plant-consuming insects, whereby every meal is the same. As such, many small farms throughout the globe practice mixed crop farming to encourage the biodiversity of beneficial insects that keep pest populations in check.

8.4.1 How Growers can Introduce Crop Diversity while Maintaining High Production Levels

Studies have shown that an effective method is to introduce genotype mixtures of one crop species having multiple nutritional levels. For instance, the parts eaten by consumers (such as corn ears or broccoli heads) may be the same; however, the parts eaten by insects (such as leaves) may be different. This will keep the products sold on the market consistently and at the same time bring variety to the crop.

Genotype mixing of this plant has been carried out in some rice and wheat fields to control the spread of destructive insects in crops. This kind of genotype mixing of crops is already being performed on some rice as well as wheat fields, for purposes of managing the spread of damaging insects among crops.

Mixed species cropping: This is the growth of two or multiple plants in one field, probably in part, throughout one year. Cultivation of mixed- species can strengthen farming systems, thereby improving overall production and insect biodiversity in crop fields. Hybrid planting is considered a promising skill for the development of practical agricultural systems, since it usually has a multi-functional role and may provide certain ecological services around the farm system, like introducing new beneficial insect species to a region they were previously absent.

For example, it can include the inclusion and recovery of organic resources, water management, and protection of the soil from corrosion and pest population suppression. This functional diversity contributes to the ecological process to provide sustainability to the entire farm system. In tropical regions, mixed intercropping methods are very common, where two or more species are mixed to promote plant and beneficial insect diversity, but it's not commonly found in temperate zone organic gardening. Even though productivity may be high, they typically require extra intensive farming systems.

Row intercropping: It involves arranging two or several crops in a row, and then ribbon planting, growing two or multiple species in a band wide enough to permit for separate management systems, but close enough to mutually influence one another, besides extensive research has recently been conducted in the European Union and the United States on row cropping because of their greater mechanization potential.

The latter two types of hybrid planting have furthermore been applied to control the system to provide other ecological services, such as providing nitrogen (N) to nourish the soil, which provides nutrients for beneficial insects. They have further been widely researched as pest management tools. And to a lesser level as a choice for disease management. Other benefits include enriched soil structure, better weed control, improved microclimate control (such as growing tall crops to offer a windbreak), creating a habitat for the natural enemies of harmful insects, and even serving as a trapping crop.

Mix seeding and insect pest attack trends: There are various studies around the effects of intercropping on insect infestation, however, they are typically contradictory since it's hard to understand the ecological aspects that may affect the relationship between insects and plants. The number of natural adversaries of destructive insect pests in intercropping is higher in 53% of researches and lesser in 9% of studies. Nevertheless, in order to use mixed plant species as a useful tool, it's essential to understand the fundamental systems involved.

The hypothesis explaining these phenomena is that insects are drawn to and focused on their food vegetation resources (easier to find or more obvious in a single (monoculture) system), while the natural enemies in the latter are more in multiple crop systems. The proper/improper landing theory is that the pests will only settle on the plant when different host plant factors (such as visual stimulus, taste and smell) are satisfied. Compared with multiple cultures, a single culture is more likely to encounter "wrong" stimulation. The intricacy of the general situation makes it hard to design crop intercropping without any practical understanding of insect pests and plant biology.

Examples of cross planting which can effectively inhibit pests are: using clover seeding to prevent cabbage root flies and using alfalfa to prevent carrot root flies. The use of trap plants, typically in band crop systems, shows the impracticality of variation. The simulation model shows that the efficacy of trapping crops is related to the comparative strength of the attractive crops to the main crops. Case in point, various green manure plants have proven to be great trapping crops for some grass in the lettuce agricultural ecosystem.

8.4.2 Why monoculture is bad for insect biodiversity?

When flying, most of the crop squares seen on the ground are of a single culture. But monocultures are not just agricultural art: they are the key to agricultural production in farming areas. However, it is interesting that they may at the same time demonstrate to be the collapse of the agricultural method as it is known. Well, monoculture refers simply to large-scale cultivation of one crop in a selected area. Over the years, scholars have talked about the impact of cultivation and harvesting acres of land with just one crop—monoculture accounts for 442 million acreages of land in the U.S (Konvalina, 2016).

The use of monoculture agricultural production systems has some advantages, specifically- convenience. Through planting a single cultivar, it can be planted, watered, and harvested easily since there no need for sorting-agricultural pest management can also be standardized and can be accomplished in a significantly short time. The benefits of a single culture seem reasonable, and in various ways have completely changed agricultural manufacturing. Upon additional investigation, the outcomes of agricultural monoculture may actually be destructive to the ecosystem that beneficial insects live in (Konvalina, 2016).

8.4.3 How Monocultures can damage Insect Biodiversity

First, a single culture damages the soil. Through planting one plant type in one soil for several years, the soil will become more worn-out over time. It's a problem in itself-because it can lead to the generation of lower value plants.

However, this will not happen since there are fertilizers and other compounds that may (synthetically) restore soil quality. With the popularity of monoculture, people generally use a lot more fertilizers (actually lots of fertilizers), which ultimately pollute the atmosphere and surrounding aquatic sources that support insect populations.

Monoculture reduces biodiversity (indirectly increasing the demand for pesticides). In the establishment of a system to prevent or eliminate biodiversity, two things occur: First, there isn't any practical organic pest management- various kinds of insects can't control their own populations or other insects' native populations-which causes the second issue: Since there isn't any natural assistance in pest management, increased pesticides are required to manage pest populations (Reddy, 2017).

In addition, monoculture stimulates pest infestation. Since monoculture is an acre of one plant, the whole monoculture can attract one kind of pests. Since there's no biodiversity for mitigating these effects, besides because pests can easily attain food and reproduce, pests can invade the whole monoculture. For coping with this intrusion, additional pesticides may have to be used which destroys insect eggs upon spreading to the surrounding area.

Similarly, plant diseases are more likely to spread. In monoculture, plant diseases are more likely to spread, because all crops in monoculture are similar, so once the disease spreads, it is difficult to stop. How can farmers stop the spreading of these diseases? Yes, using more pesticides. As said here, the invasion of pesticides and fertilizers used for maintaining a single culture is extremely destructive to pollinators. When bees or other pollinators become exposed to pesticides (particularly new pesticides), they are more vulnerable to other ecological stresses, which are usually immune to them.

In addition, the reduction in the number of pollinators caused by new pesticides translates to a reduction in pollination, which may eventually cause a change in agricultural varieties. Besides, the imbalance of the ecological balance triggered by the reduction of pollinators has affected biodiversity and other ecosystems, apart from destroying the food chain of insects. Getting rid of monoculture may greatly reduce the demand for pesticides, and therefore environmental contamination (Reddy, 2017).

Improper crop rotation and the lack of nutrients in soils have caused a series of problems, which have also prompted people to get rid of monoculture. Agriculture is a type of culture. But other factors (whether prosperity or possible extinction) of biodiversity are needed to understand it, this is not a fragmented activity: for instance, resources dedicated to crop herbivores may increase the area accessible for pest migration.

This simplification also reduces environmental prospects for natural enemies. Therefore, when a large number of immigrant pests and beneficial insect populations are suppressed, favorable weather and fragile crop growth stages occur at the same time, and pest outbreaks often occur.

The anti-insecticide campaign must be accompanied with a concentration away from damaging monocultures, which requires improved use of pesticides, which not only reduces the use of systemic pesticides but also reduces the demand for systemic pesticides. The campaign should also comprise a preference for organic farming methods and a focus on long-term solutions that devote resources to pest management, which can supplement ecosystems instead of destroying them (Shields, 2017).

8.4.4 Mixed-crop Fields can Enhance Biocontrol Measures for Invasive Insects

Figure 8.4: Mixed crop fields enrich the environment and provide nourishment for different types of insects.

Source: https://www.americanmeadows.com/blog/2016/02/17/how-to-prevent-naturally-treat-harmful-insects-in-the-garden

The latest research from China Agricultural University investigates how the richness of plant species in farmland affects insect communities. In a test field with multiple crops growing at the same time, scientists discovered that more plant species in the field were associated with more insect types present there. In addition, as the varieties of plants in the plot increase, the fraction of natural enemies for pests also increases (New, 2012).

Contemporary intensive agriculture has a wide field of single- crop cultivation of economic crops, which seems to be very rich with regards to output and profits, but in the eyes of ecologists, these fields are quite barren. With regard to species richness, they are quite poor.

Spccics richness basically is a measurement of biodiversity, covering only the amount of species in existence. It doesn't factor their richness ,however. Carefully managed, weed-free wheat is essentially rich in plant species richness, while native grassland areas of similar size might have multiple plant species.

Agroecology is a study on how to apply ecological principles to agricultural practice. For example, the ecological variety-stability theory mentions that the higher the diversity of plants in a given ecosystem, the more secure it will be. Some agricultural ecologists believe that increasing the diversity of plants in farmland through polyculture (growing multiple crops at the same time in a field) can make production more stable plus farms more viable. However, how does polyculture that increases the richness and diversity of plant species impact the diversity of insects (including pests) living in the field?

8.4.5 Case Study

In a research published in the American Entomological Society Report, scientists from the China Agricultural University and colleagues from China, the U.S., and South Africa tested if polyculture can help farmers to manage pests found in the fields.

They studied how the richness of plant species in farmlands affects insect communities. The scientists divided insect communities into three categories: pests, which destroy plants through consuming them; natural adversaries, insects that feed on plant pests; plus, environmentally friendly insects that are harmless or do not provide pollinating and other beneficial services.

The researchers sought to know whether the richness of plant species would affect these insects differently. Will increasing plant species diversity in farmland stabilize insect communities? Will it increase the efficacy of natural enemies and help them regulate crop pests? To discover more, the team developed an ambitious experimental model structure, which is essentially a tiny farm with around 50 small fields and plots measuring (9 meters x 9 meters), with each farm randomly grown with 1,2,4,8 or 16 plants which are chosen from 20 diverse crops.

After that, within three years scientists sampled insects found on the plot and calculated the biomass of plants and insects. This is a difficult task. Every plot was planted with about 484 plants and weeded manually (without applying any pesticides or herbicides). In addition, more than 250 kinds of insects were found during the entire study.

Scientists have discovered that the higher the plant species found in a given plot, the greater the number of insect species present there. The same is true when looking at the three types of insect groups, pests, natural adversaries and environmentally beneficial insects. Greater plant species richness in each insect category is equally associated with greater insect species richness and higher overall insect biomass.

In other terms, increased plant species support additional insect individuals. However, the impact of flora species abundance on biomass is not the same in all insect categories. The greater the plant species available in this land, the greater the biomass capacity of natural enemies, however, the richness of plant species has no substantial impact on the pests biomass and environmentally beneficial insects. This ratio indicates that natural enemies may provide valued pest management tools for biological control services. If there are additional predators around, they may stop the surge in pest numbers.

Asian multicolor lady beetle: In a study on the relationship between crop species abundance and insect biodiversity in ecosystems in China, the Asian colorful beetle (*Harmonia axyridis*) is among the natural hostile insects that appear in large amounts in mixed cropland. It is also related to a higher ratio of natural enemy pests. Researchers mention that when there are not many pests, possibly environmentally friendly insects may be applied as a substitute for predators. Having alternative food resources can stabilize the natural enemy population and maintain its high number. The researchers determined that increased plant species abundance in the field may support insect communities and improve the biological control of natural enemies against crop pests. This is micro-landscape research, which revealed the effects of spatial designs on biological control factors at a great scale.

8.5 CROP ROTATION FOR MANAGING INSECT POPULATIONS

The effectiveness of rotation as an insect management tool relies on the lifecycle of target insects. For crop rotation technique to control pests appropriately, insects should spend time from the finish of one crop up to the start of the subsequent crop in a lesser-mobility stage, plus the variety of host crops should be restricted. Few pests fit this model. The majority have a phase in adulthood, and they can simply cross at least a farm.

Generally, this highly active phase occurs when insects overwinter in spring, so rotation from a year to the subsequent won't affect them. A number of pests, such as cabbage loopers, potato leafhoppers and corn earthworms, don't actually overwinter in the U.S. northeast (Kumar, Yogi, &Jagdish, 2013).

Instead, they move hundreds of miles every year to re-infect the host crop. Nevertheless, certain pests can be controlled by rotation. The Western corn rootworms as well as northern corn rootworms make great examples. Adult beetles consume corn silk (including many flowers, such as the cucurbit flowers) which bloom usually around August and September while laying eggs in soil within the corn plant base.

Typically, the eggs overwinter before the newly hatched larvae consume corn roots during the spring season. They cannot survive on tiny grains, sorghum, broad-leaved crops or weeds, but only on grass weeds at rather low levels. Therefore, these two beetle types rely on extensive planting of recurrent corn to preserve their survival, plus only when continuous corn planting has become a widespread activity in the Midwest is it a major pest.

Similarly, due to the continuous cultivation of potatoes, especially in the northeastern United States, the Colorado potato beetle has become a serious potato pest. Beetles appear in late summer (in Massachusetts in August), and were mainly on the edge of the field in adulthood, but in the soil, they also matured within the same place of their development.

When they appear after the spring overwintering in the second year, they prefer walking to the host (if available in the area) and there will be a delay before they recover their flying muscles. By moving the potatoes to another field away from the preceding year's potatoes and using various obstacles (plastic-lined ditches, intensive planting of cereal crops, etc.) to inhibit the spring infestations, infestations can be reduced in the new area.

The life cycle comprises the long-distance flight phase of the first generation of adults, which appears in the middle of summer and can hit fields not far from the preceding year's potatoes, however, rotation is an effectual way to prevent early season loss (Kumar, Yogi, &Jagdish, 2013).

Wireworms (click beetles larvae) and hairy genus plants are local pests. They can accumulate in the field behind turf or small particles and then stay for many years because the larvae live long. If monitoring indicates problems, root crops or corn must not be planted within the field.

8.6 IMPACT OF ZERO-TILL FARMING, SEASONAL WEEDS AND CROP RESIDUES ON INSECT PESTS

Crop rotation plans must take into account the entire field needs and operating range. Field management practices may affect pest problems, so the entire rotation plan must take this factor into notice. The research literature, mainly written on corn and other types of field crops, shows that no-till farming presents a mixed impact on different types of insect pests found in a particular field.

Due to crop residues, cool and moist soil conditions, and no-till tillage, the disturbance to the soil is small, and some low-activity pests, especially black worms, Spodopterafrugiperda, slugs and nematodes are favored. However, similar conditions in the field also benefit the natural enemies of many soil pests, such as beetles, predatory mites and spiders on the ground. In landscapes composed of large fields cultivated at least yearly, undisturbed vegetation bands around the edges of the fields are important sanctuaries for these predators and parasites of insect pests like the European corn borer, Spodopterafrugiperda, and grain leaf beetles.

Migratory pests might overwinter among crop residues within the field, so cultivating and decomposing this residue may diminish their survival rate. The European corn borer, pumpkin vine bore, onion worm, and spinach leaf borer are all illustrations of insects in this category. Their overwintering stage and the remaining crop residues can be damaged by cultivation. Since these species are usually adults when they appear during the spring season, the impact of greater mortality rates might not be obvious in one field or farm, however, timely cultivation in the entire area may diminish the number of local populations.

Tillage also destroys yearly or biennial weeds during winter, which may attract pests to the field at the start of the season. Furthermore, black scale moths produce their eggs during early spring and favor lowland, poorly drained, weed -filled fields. Caterpillars start to grow on weeds or crops. When these host plants are cultivated and swapped with cash crops, the caterpillars will attack the stems of their new crop unless sufficient time of between (10 days to 2 weeks) has been set aside to starve them to their demise.

Before the emergence of other host plants, cruciferous weeds, particularly yellow rockets, likewise attract cruciferous flea beetles (Phyllotretacruciferae) that enter the field at the beginning of the season. Flea beetles overwinter in adulthood and become active, starting to forage from the starting warm days of early spring. The newly plowed residue is still decomposing during sowing, attracting seed corn flies to produce eggs. Besides, the larvae hatch and consume sprouted corn, beans or gourd seeds, or channels into the stems of fresh grafts.

8.6.1 Planting Interval and Insect Pests

Some pests, such as the corn flea beetles, striped flea beetles, crucifer flea beetles, as well as cabbage, caused most of the damage to small seedlings early during the season. Damage may be prevented by inhibiting planting or using row covers to protect early planting. A New York State study formulated a rule to avoid cabbage worm damage by waiting for seedling transplants of maturing cabbages until the yellow rocket part reached its peak.

The time to avoid damage to the flea beetles by the crop can delay planting up to mid-June in places like Connecticut. Different other pests, like the Mexican beetle, emerge from the overwintering site (around late June) but prefer laying eggs in the biggest and most dynamic cultivations, which are usually the earliest. Biological regulation programs through the parasite Pediobiusfoveolatus (a tiny beneficial wasp) have used this preference to plant early trap plants on attracting host plants (snap beans), and on these Wasps were released from the early plantations that were severely infested.

8.6.2 Movement between Crops

A good rotation plan comprises not just the sequence of plants to plants, but also how to arrange crops on the farm throughout the season. For some specialized pests (such as cruciferous flea beetles) for many generations each year, continuous planting of a similar crop in the field can increase the chances of establishing a high population. Although adults usually have a spreading stage that is easy to spread over long distances, even in a life stage that usually only spreads a few yards, placing multiple plants next to each other can help insects find fresh food plants.

For many specialized pests (such as cruciferous flea beetles) for many generations each year, continuous planting of one crop in the field will increase the chance of establishing a high population. Other pests feed on various different host plants plus may accumulate in large amounts on one host and then move to another host. Understanding these possibilities can help growers avoid potential problems.

For example, tarnished insects feed on hundreds of broad-leaved host plants, choosing to forage on flowers and nurturing seeds. Nymphs may grow on hairy purple etch plants, weeds (such as weeds or legumes) or wayside vegetation. The new adults later move out once the plants have been mowed, or after the seeds become mature and less convenient for them to stay.

Strawberry growers who are trying to lessen harm can exchange rye (a host of moldy plant bugs that have deteriorated) with rye and a blend of rye and shaggy purple etch vegetables (favorable hosts). Strawberry growers that include alfalfa and clover in crop rotations can furthermore avoid pruning of these legumes during strawberry flowering and fruit development, thereby avoiding the transmission of speckled adult plant insects.

Usually, some major pests can be directly controlled by rotating crops on one farm, however, more are affected by farm management practices, such as if and when to cultivate, when to plant, managing spring weeds and if it's appropriate to plant in one field continuously. Because of the overwintering of insects, the distribution and host discovery strategies are so diverse, it is difficult to detect cultural practices which can inhibit problems with various crops and pests.

The best method is identifying the most common pests that have traditionally destroyed major crops on a specific farm, and discover cultural procedures that interfere with the lifecycle, movement, and even restrain the population of pests or destruction caused by those pests.

8.6.3 Effects of Mixed Planting on Population Trends of Insect Pests

In many countries, the mixed or intercropping of plant varieties and cultivars is quite a popular cultural practice. One of its possible advantages is the impact on the populace dynamics of pests, which can minimize damage to crops. This chapter mentions some of the common effects related to the initial planting of crops and the nourishing, reproduction, mortality, and spread of pests in the crops. In most developing nations and portions of some developed nations, mixed or intercropping plants is a widespread cultural practice. In this method, two or additional crops or wild plants can be simultaneously planted in one field.

Differences on this theme comprise of mixed intercropping (no obvious row configuration), intercropping (growing one or additional crops in a row), intercropping crops (crops growing in different zones, enough to allow independent cultivation) plus relay intercropping (in each crop two or additional crops developed simultaneously during some part of the whole lifecycle of the crop, apart from planting the second crop before harvesting the first crop).

The potential benefits of intercropping include the impact on pest population dynamics, which may reduce the destruction of crops. For instance, in northern Nigeria, cow beans grown as a single cultivar will hardly harvest any crops unless sprayed by protective pesticides, while untreated cow beans in mixed planting systems can produce valuable harvests and sometimes the use of a single pesticide produces more responses than single cultivation.

Cross planting can especially affect crop colonization as well as successive population growth and survival. For crop planting, the dense leaf crowns related to mixed planting are less visually appealing to some insects, and these insects have a visual reaction to plants with wide ground spaced against the exposed ground, which is notable for certain agricultural ecosystems.

Although on the contrary, fruit flies and sorghum sprouts previously associated with wild grasses may be easier to colonize in dense areas than sparse lands. High crops like traditional cultivars such as millet and sorghum might become physical obstacles or traps for insect vermin that invade from outside the cultivation area, or even strong aromatic plants like onion, lemongrass, tomato and garlic can provide an olfactory disguise for other susceptible crops like vegetables and grain legumes.

If a particular crop in the mixture is in a more appealing growth stage compared to other crops when the pest invades, it is possible to transfer harmful organisms from adjacent crops, even though this is hard to attain in practice. Therefore, if the insect attracting ear of corn coincides with the time of bud formation on cotton crops, corn will attract Helio moths on cotton, otherwise, corn is usually the main source of harmful populations of Helio moths on cotton. The use of trapped crops is especially attractive because it can concentrate invasive pests at the beginning of the season, so pesticide application can be restricted to eradicate them.

8.6.4 Population Growth and Survival

If potential pests successfully color-crop the plant, proper intercropping may reduce their development and survival rate, thereby minimizing the possibility of their population exceeding the economic damage threshold. Where the host and non-host mature together, especially where the intercropping space is larger than normal, it may hinder the activities of live pests and airborne pests in the soil.

In Malawi, researchers noticed that the hook hairs on kidney beans captured scattered aphids and efficiently reduced the rosette viral infection of nearby peanuts. Regrettably, some pests, like the grasshopper Mel Melanoplusbivittatus, produce eggs at the verge of the production zone, plus are more intense when producing eggs at the verge of the crop. Strip cutting surges the fraction of edges in every cutting unit. At this time, it's not yet determined whether there are additional beneficial species than only two species in restraining the dispersal of pests, or the fraction of non-hosts needed to improve the protection of hosts which are endangered.

Limiting the spread may also be due to the mixed-resistant and susceptible varieties of a crop, with belief that the main process of reducing pest problems in mixed culture is: (I) disperse individual losses by dispersing on non-host components ("flying paper effect") and (II) Compensatory growth of components that are not attacked, which will enhance the flying paper effect.

The prior is likely to work in situations where the proliferation stage of a pest insect's lifecycle comprises passive transportation and random, undirected motion so that the capacity of individuals exposed to a susceptible host relative to those exposed to a non-host depends on the comparative abundance of exposed matter.

Diverse non-host and resistant crops are responsible for curbing infection in the mixture, suggesting that the reduction in disease transmission is largely a matter of the density of susceptible plants, at least once the fraction of resistant crops is minimal, and therefore pathogens have fairly minor impacts on these situations.

A team of scientists proposed a mathematical framework of a more complicated situation, which includes a nematode found in a dual-mode crop system, whereby the parasites selectively attack the material with greater yield capacity. This model is used for predicting the long-term impacts of strategies that involve pure-breed and /or mixed crop rotation on the field.

When there is no inoculum migration, the best technique is rotating the pure culture, and in the case of a fairly strong flypaper effect, the periodic introduction of inoculum is best solved by mixed planting. Unfortunately, there are few experimental data on the dynamics of insect pests in mixed host populations, including the proportion and spatial distribution of vulnerable and resilient lines that provide the greatest protection.

The results show that transferring the benefits of mixed planting to monoculture varies from place to place. In many nations, mixed planting will undoubtedly continue for a considerable period of time, and research must aim at understanding and retaining the inherent "natural control" of these systems and combining it with appropriate chemical regulation. In some cases, including many temperate lands, it is essential to transfer the benefits of mixed crops to agronomically appropriate monocultures. It's been revealed that mixed cultivars may sometimes attain this purpose, but there are more possibilities, especially in mixtures, where the key factors that provide insect pest protection are chemical in nature.

Nevertheless, in circumstances where complex general insect predators aren't as important as a few particular natural enemy species, interspecies may cause destructive interference and disrupt the pest and natural enemy balance. As for the ingestion and reproduction of pests, confused olfactory and optical stimuli from the host or non-host may disturb normal eating or mating trends, and the intensive cultivation of intercropping crops may cause a decline in the nutritional value of crops.

To conclude, there is no doubt that components of natural pest management exist in various mixed crop systems. Although chemical elements in non-hosts that help protect crops in the mixed field are easier to transfer to monoculture compared to physical factors (such as camouflage), they can sometimes be modified enough to be adopted in modern temperate cultivation methods. An illustration of this is the effect of weeds growing on Brassica fields on the optical responses of aphids and egg- laying habits of cabbage root flies.

Furthermore, given the rising cost of discovering and retailing synthetic pesticides, there may be reasons to use more alternative methods, like beneficial insects, and plant extracts as natural repellents and feed or propagation inhibitors, especially if these extracts are by-products of the crop processing industry.

9

Challenges Facing Insect Conservation

CONTENTS

9.1 INTRODUCTION

We are living in a world that is rapidly changing. Yet, we do not pay attention to the fact that the major aspects that we depend on are in jeopardy. One of the main aspects of these is insects. Despite the fact that they are tiny and hardly ever seen as they perform their various important activities, they are very necessary in the maintenance of the world in general. Insects normally have so many different tasks they perform, that if at any one point they go extinct, it would be a huge catastrophe. It is very sad that very few people actually understand how important they are in our daily lives.

We need to understand that insects are probably the most dominant living creatures on earth. This makes it sadder because many human activities are threatening insects populations. Yet, there is very little time for us to fix this problem of increasing insect loss all over the world.

Despite the fact that there are many strategies already in place for dealing with this issue, there are some that have proven to be more important in ensuring that we successfully conserve insects into the 21st century and even beyond. So as to understand the importance of insects, we will discuss some of them:

Figure 9.1: Pollinators are very important in the maintenance of our natural ecosystem. In fact, Approximately 1,000 of the plants grown for medicines, fibers, beverages and food, need to undergo pollination by insects so as to produce the food which we rely on.

Source: https://www.wildlifehc.org/the-importance-of-pollinators/

- Pollination – the value that insects provide when it comes to the pollination of plants cannot be calculated. Amongst the most important pollinators, are honeybees, and the work they do results in an estimate of 80% of the total pollination in the US. In the United States, honeybee pollution has an impact on approximately 20 billion dollars every year in crops alone, including nuts, vegetables, and fruits.
- Foods – high on the list of foods we get from insects is honey. Honey is highly consumed in all parts of the world by humans. There are some people who eat insects as novelties in the US, however, they are used as foods by other societies, such as beetle grubs.
- For Silk – it was back in 2640 B.C. In China, when silk was recognized as a highly valuable product. As of today, China produces approximately 31,000 tons of raw silk every year, accounting for 80% of the supply globally. The majority of the silk produced is from silkworm cocoons mainly *Bombyx mori*.
- Biological and Natural control – nature's balance is highly dependent on the works of predators and parasites, many of them being insect species. This concept is used by researchers in biological control, and this has apparently been quite successful in many different programs.
- As aesthetics – in many different areas of art, insects are well known as being highly pleasant sensory-wise. In nature, the most appealing creatures are butterflies, having both patterns and colors that humans enjoy at different times of the year. Many societies have used insects all through history, with no limitations large and/or colorful beetles and butterflies. In the United States, the Native Americans utilized insect parts in a way that feathers were used for crafts. In Ecuador, the Jivaro Indians were able to make earrings from the brightly colored wing covers of specific beetle species. The scarab beetles were used by the Egyptians as their sun God symbol. On ancient Greek coins, bees were depicted. Many different art branches have been able to exhibit insects in one way or another, such as various selections of postage stamps used all over the world.
- For their products – some of the products we get from insects include:

- *Lac* – this product is normally gotten from Lacciferlacce which is a Lac Scale insect. The majority of it is made in India, and this is where the world gets approximately 50 million pounds of lac from every year. Many different items require lac as an important ingredient, such as varnish, printing inks, various sealants, insulators, shoe polishes, and floor polishes.
- *Beeswax* – 1 million pounds of beeswax is imported every year by Britain alone. It can be utilized as a base for candle making, polishes, and ointments. 40% of the total produced beeswax is utilized in the manufacturing of cosmetics for lipsticks, creams, and lotions.
- *Dyes* – there are many different scale insect species that provide dyes which are used in various products, such as beverages, medicines, for coloring cakes and cosmetics. The bodies of scale insects, specifically *Coccus cacti*, produces cochineal which is a bright red pigment. During the first decade of the 20th century, certain synthetic colors were competitors with the insect dyes, however, these dyes were later found to be carcinogenic. This means that the insect dyes ended up flourishing in the end. Insect galls can be used to produce tannin, which is used to tan hides, as well as in producing permanent durable inks. Other dyes are produced by gall.
- *Genetics* – there are many different fly species, specifically fruit flies, that have been used in the study of genetics, and they tend to be very efficient due to the fact that they have a short lifespan of approximately ten days.
- *For dermestids used to clean skeletons* – the carpet beetle is a tiny insect that feeds on the majority of the organic products, such as dried insects in collections, carpets, and cereals. This fact is taken advantage of by museum technicians, utilizing dermestids colonies that are established for cleaning mammal skeletons.

When you think about the insects that we have talked about, we are able to see the different threats that these insects face. In general, when put together with terrestrial invertebrates, we can state the major threats as:

- Loss of habitat due to logging and agriculture.
- Development of infrastructure such as urbanization.
- Fragmentation and loss of habitat due to transportation or service corridors.
- Invasion by alien species.
- Changes in the fire regime.
- Pollution.
- Severe weather and climate change.
- Mining.

When it comes specifically to fresh water invertebrates, dams and poor management and pollution are the main threats, and these are much greater than when you put all the others together. We need to understand these threats so as to understand how they are affecting the conservation of insects.

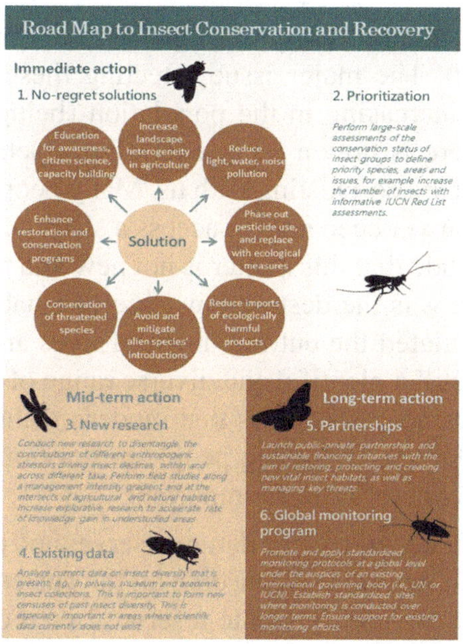

Figure 9.2: Above are some of the aspects we need to take into consideration for the conservation and recovery of insect species. The immediate actions include no-regret solutions such as avoiding and mitigating alien species introduction, reducing light, noise and water pollution, phasing out the use of pesticides, etc.

Source: https://news.wsu.edu/2020/01/09/wsu-researcher-joins-international-call-halt-massive-insect-decline/

9.2 LOSS OF HABITAT

This has proven to be the greatest problem in the conservation of insects. There is a major decline of insects and other invertebrate species in locations where they were previously flourishing. There is a major reason as to why we need to be highly concerned about this matter, mainly because we as humans are continuously destroying these habitats. Conservation of these habitats then remains as the main solution, however, this has proven to be more difficult than is thought of, mainly because we destroy these habitats as we look for prosperity, housing and food. There may be a price to pay if we choose areas for insects to live. Some of the prices we may have to pay to include the diversion of a new road away from a conserved valuable location or even lower agricultural yield. In addition, insects play an important role in the existence of human beings, and we will be greatly affected if we do not make sure that their habitats are conserved. In addition, there are several actions we can take at hardly any cost to ensure this conservation.

As far back as 400 million years ago, insects thrived perfectly well without us (humans). The major issue affecting these habitats is the fact that we are greatly increasing in the population (being approximately 7.6 billion) and thus there is now a high level of competition with the many different species that are co-existing with us. Many people all over the world are poor and have not yet been able to access a share of the resources in the world. We can conclude that this matter is not new and thus not a new crisis. It is a fact that there was the destruction of certain habitats by our ancient ancestors as they initiated the cutting down of trees and tilling of the soil, however, there was still a plentiful and diverse range of habitats on the land. The last few decades have shown us that modern technology has been the main driver of habitat loss, allowing people to develop changes on a scale that is now making conservation so important.

It is not enough to depend on the nature reserves and other areas meant for protecting wildlife. If we want to conserve all the currently existing species, these are not enough. When we create reserves based on the fact that most of the earth's resources are meant for us, meaning that we need to set aside small portions of them as wild reserves, we limit diversity. Theoretically, if these reserves and other areas that are protected could be a representation of all the different varieties of habitats that are found in one specific region, there is a possibility in supporting all the region's species. Sadly, however, it is not practical to set up a range of reserves that represent all species.

Even if it were possible to fully represent all species in protected sites, it would not be a success. This is mainly because, in nature, habitats hardly ever remain unchanged forever in the same location. For example, changes in vegetation can lead to making the local area unsuitable for the flourishing of insects.

Even in different years, weather fluctuations can lead to a period of temporary unsuitability for certain insect species, regardless of the habitats remaining excellent. When a locality is affected by adverse changes, it could lead to the extinction of the species within the specific area, unless there is a different locality close by that is suitable enough for the survival of the dispersing individuals. There may be a number of suitable localities close to each other in a large reserve, however, when it comes to smaller reserves which are further from other suitable sites, there may be loss of species over time. If not managed for diversity, even a large reserve can end up losing species. The only way to deal with this is by not only using protected areas but also, at any place possible, within the towns, forests, farms and other landscapes we see every day. This helps in maintaining the network of the populations shifting from one place to another.

It is also important to understand that no matter how much we manage protected areas, it does not eliminate the need to ensure the conservation of habitats all through the landscape. If we are aware of the exact requirements of habitats for different species of insects, we would be able to create or maintain the proper conditions for the specific species by making sure its habitat is gardened. The major problem with this problem is that we only know very little information about very few insect species all over the world. There is also a much bigger problem in this situation: when we 'garden' for the benefit of just a few species, or only even one, we may end up harming the rest and probably even drive them to extinction.

Figure 9.3: Insects are facing a calamitous loss of habitat. This is one of the major challenges of insect conservation, mainly because of the growing population that is taking down insect habitats to modernize or urbanize the world. This is driving away many important insects that are necessary even in the urban areas.

Source: https://climatenewsnetwork.net/insects-face-calamitous-habitat-loss/

It is possible to manage a nature reserve on a good basis through knowing as much as one can about the preferred lifestyle kind that has supported or previously supported insect species. Despite the fact that we only know the precise requirements of a few insect species, we are still able to recognize important features of the habitat that are required by various species that have the same needs. For example, there has been research showing that various grassland insects require a mosaic of tall and short vegetation at the various life cycle stages. This has also proven to be important for many different species that have not yet been discovered or studied. If there was the management of sites that only require long vegetation (or short vegetation), others could end up dying out.

It is often in protected areas that habitats are managed, and these are not nature reserves, however, their management is mainly for different reasons. An example includes the SSSI (Sites of Special Scientific Interests) in the United Kingdom, where forestry, farming and other economic uses are permitted, subjected to rules that are meant to ensure the protection of the biological and also different site values. The sites have helped in the maintenance of insect species meta-populations, however, it has proven to be necessary to ensure that habitats in the surrounding lands are conserved.

There is a lot of work that needs to be done in the conservation of habitats. All over the world habitats are damaged, making it more difficult to deal with the different situations in which it happens. In the existence of natural ecosystems, for example in the tropical rain forests, it is important to ensure that they are left untouched. This tends to seem impossible, however, countries in the world that are richer should make efforts at helping poorer countries in affording to slow down or halt the destruction rate.

Despite the fact that there are many species of insects dependent on natural ecosystems, there are also many others that can live with us on lands that we utilize for crop growing, or even around industrial sites, along roadsides, and around our houses. This is inclusive of all the species that are found these days in countries such as the United Kingdom, where there has been virtual management of land surfaces in one way or another. There is now a decline in these species due to the fact that they are undergoing deprivation of habitats that persisted or developed alongside our ancestors.

Below is a list of a number of things that can be done for the conservation of insect habitats and other invertebrates in the United Kingdom as well as various countries (developed and developing):

- *Farmlands* – we should aim at not destroying dry stone walls or hedgerows for the amalgamation of fields unless it is an important task. We should also avoid cutting both sides of the hedge in one year; this is very helpful in ensuring that the insects survive for much longer in the arable fields.
- *Arable land* – we should also refrain from the use of fertilizers and pesticides around the corners and edges of the fields as well as close to the hedgerows. If the arable land is already very huge, the creation of strips that are uncultivated (beetle banks) across them, for the restoration of habitats as well as for encouraging natural enemies of pest species, is necessary. The habitat potential of land that has been set aside; for example, through the provision of many different insect food plants, as well as through the utilization of as little herbicides as possible.
- *Grassland* – grasslands should be retained as much as possible. When one is controlling conservation grassland scrubs, it is important to maintain a certain proportion of woody plants by allowing them to remain, even if it means leaving intreated stumps without using herbicides. If there was fertilization and reseeding of the grassland, it is recommended, if possible, to restore a herb-rich sward. For the creation of many different sward heights, it is possible to use control grazing; avoiding both under-grazing and over-grazing. This can also be achieved with the use of traditional

breeds of livestock that are suitable to the local conditions. If it is feasible economically, it is important to try reducing or stopping the use of artificially made fertilizers and also trying to convert pasture that has been improved to a condition that is more herb-rich. If there are ancient trees in a grassland, for example in a pasture woodland, it is important to ensure that majority of the trees' rooting areas are protected as much as possible from livestock trampling on them. It is also necessary to restrict or avoid the use of veterinary pharmaceuticals, for example ,Avermectins, which makes livestock dung toxic to invertebrates that feed in dung.

- *Deadwood and woodland habitats* – a proportion of trees should be left to decay and age in a natural manner, instead of harvesting all of them. Woodland rides should be kept sunny and open, with a shrub layer edging. It is also advisable to create glades through coppicing periodically, however, not in coppice neglected areas that have ended up becoming important for insects depending on sap-runs or deadwood. Protection of veteran and ancient trees (the majority of which are found in pasture woodland, instead of high forests), as well as trying to make sure that their places are taken by younger trees. Firewood piles should not be left in areas (for example in wood pastures or ancient woodlands) where they end up acting as decoys for deadwood invertebrates that are rare. In schemes for farm woodland, it is important to avoid planting in places that are habitats good enough for grassland or wetland invertebrates.
- *Mountains, moorland, and heath* – stands of heather should be maintained through periodically burning or cutting of small areas, however, large areas should not be burnt. If open areas are turning into woodland or scrub, some of the scrub or trees should be cut down so as to ensure the protection of the open habitat, however, some should always be left for habitat diversity and shelter. Larger-diameter cut stems should be retained, especially for deadwood habitat. Strips and patches of bare ground need to be available all through the area, for example avoiding path improvement, where this could lead to the reduction of the amount of bare ground habitat.

Figure 9.4: We are highly dependent on animals, insects and plants for the delivery of natural pest control, crop pollination, flood prevention, soil conservation, and water quality. We are now threatening all this, not only due to the growth of cities and clearing if forests, but also through the excessive use of fossil fuels releasing greenhouse gases to the atmosphere, such as carbon dioxide, which is leading to global warming.

Source: https://www.mercurynews.com/2018/11/09/letter-could-insect-die-off-be-our-climate-canary/amp/

- *Water margin and aquatic habitats* – avoid straightening or deepening rivers and streams; schemes should be supported so as to make sure that such changes are reversed in situations where this has already happened. If there has been an artificial deepening of water bodies, gentle shelving edges should be restored as much as possible. Projects for the protection or restoration of natural processes should be supported, such as the formation of dynamic sand in river beds and banks of the shingle. The maintenance of open vegetation, such as reed-beds found in marshy areas, while retaining some of the woody plants such as alder and willows for the uniquely provided habitats is important. Fallen timber should not be removed from watercourses, and this can only happen if it poses a major risk to property or people due to potential flooding. So as to ensure recovery of fauna and flora, it is important to restore ditches and ponds that are seriously silted up; this work should be done in stages for the recovery of the fauna and flora. There should also be no building of any things on the flood plains, as this causes the destruction of habitats because of flood defenses.

- *Gardens and parks* – some 'wild' areas should be provided. If a tidy and formal appearance is necessary, it is possible to achieve this through the maintenance of flowerbeds and neat lawns, but only in specific sections such as along paths. Instead of getting rid of garden waste, it is much better to keep a compost heap. It is also helpful to use plants that help in providing habitats for many different habitats; not just those that are a source of nectar. If there is enough space, it is also advisable to create deadwood habitats and water habitats, and if at any point a tree has to be cut down (due to safety reasons), a tall stump should be kept. It is important to avoid the development of domestic gardens for infill housing.
- *Brownfield sites* – an example of this is derelict industrial land. Seeing as these sites may end up having a lot of biodiversities, similar to that seen in ancient woodland, the myth that redeveloping them is better than building on greenfield land should be dispelled. If there has to be the development of brownfield land, for new industries or housing, it is important to make sure that there is the provision of habitat areas in the plans, to ensure they are retained and well managed after development.

9.3 DEVELOPMENT OF INFRASTRUCTURE

During the last century, there has been rapid worldwide urbanization, leading to more than half of the population, globally, living in urban areas. There are certain areas in which studies have focused on when it comes to the effects of urbanization on the diversity of insects, such as ecosystem services; food habits, extinction, distribution and insect abundance. There has been a decline in the native insect population, and studies of their spatial distribution have been able to reveal that the abundance of insects is reduced together with what is referred to as the rural-city center gradient, and thus there is an extinction of many native insects with the process of urbanization.

Majority if the specialist insect communities have reduced abundantly mainly because of urbanization, whereas a number of generalist species, for example, termites, cockroaches and aphids, have had a significant increase in abundance. In comparison to predator insects, it has been seen that flower-visiting, saprophagous, parasitic and herbivorous insects tend to be influenced more negatively by urbanization. This has proven to provide an effect on the ecosystem service insects that is highly significant. Some of the factors that attribute to the decline in insect populations mainly because of urbanization are occurrence of introduced invasive species, clustering of buildings, road hardening, habitat fragmentation, and environmental pollution (including heat pollution, light pollution, water pollution, and air pollution).

With the continuation of urbanization, there needs to be measures that help in protecting insects found in the urban areas. This involves the improvement of basic scientific research on the issues affecting insect populations in urban areas, ensuring that the public are well informed on the benefits of protecting the environment, and constructing habitats that are suitable for the survival of insects.

Figure 9.5: The majority of the phytophagous insects are well known by the various different food items they feed on, which allows them to redistribute the trophic load to other different types of food in situations where there is a decrease in the supply of any particular component of their diet. This is however is inhibited by urbanization, that pushing insects away from these areas.

Source: http://scienceovereverything.com/2018/01/25/buzz-urbanization-pollination/

One of the main characteristics of industrialization includes moving from a mode of living that is agrarian, to one that has dominant technological innovations. In general, there are many advantages that come with industrialization, all of which have helped us as humans gain progress and enjoy various efficiencies. However, regardless of these advantages, industrialization has lead to many other problems such as global warming, environmental degrees, and pollution. Apart from humans, the detrimental effects of industrialization are being experienced by insects as well. Below is a discussion of some of the effects they are experiencing.

- *Disruption of the ecosystem* – one of the major contributors to pollution is industrialization. In accordance with a report released by the United Nations Food and Agriculture Organization, an approximate of 6.4 billion tons of carbon dioxide to the air every year is contributed by industrialization. Many of the insects that depend on the marine areas are not able to survive in these levels of pollution and are therefore going extinct. Many insects depend on flora and fauna and these are also affected by the increase in water and atmospheric pollution. When there is a threat to the insect source of food, many of these insects struggle to find their source of food somas to survive.
- *Extinction* – in accordance with the survey from the United States Geology department, 2/3rd of the insect population will not be existent by 2040. However, some insect species face a higher rate of extinction than others; such as honey bees, wasps, beetles, butterflies, and moths. Insect extinction can be highly attributed to the rise in the levels of deforestation which is done to create space for human settlement and large scale industrial farming. It is also possible for this loss in insect populations to be attributed to increased global warming.
- *Loss of natural habitats* – assisted migration or managed relocation is both new aspects that affect the animal kingdom. The two phrases are used to refer to the transference of insects from their natural habitats to different habitats in different locations. This process is mainly done for the protection of insects from pollution and extinction. This process was opposed to by scientists who were arguing that this could lead to a lack of biodiversity and survival of the insect species due to their needs and requirements that may not be well understood. In addition, when insects are

relocated, they are forced to adapt to new environments which may be difficult.

- *Insect-human conflict* – when cities are expanded, technological advances are propelled, thus requiring more and more space for humans to live in. This is one of the main characteristics of urbanization. Despite the fact that the growth of cities may deal with a growth in the economy, it may also cause land encroachment to be inhabited by insects. Due to this, there us a decrease in the insect natural habitat, forcing the insects to compete for food and space with human beings. This leads to killing the insects (such as beetles, weevils, moths, etc.) so as to lower the impact that they bring upon the encroached area, thus leading to extinction and disruption on their ecosystem and natural environment.

9.4 INVASIVE INSECT SPECIES

Invasive insect species are those that live in habitats that are not their native homes, and end up in those locations due to introduction. These insect species end up posing threats to the locality in which they were introduced, damaging the economy, the environment, and also human health. There are many different invasive insects all over the world. Their rapid spread is causing major economic and ecological impacts in various countries. Agricultural organizations and bodies that deal with the conservation of these environments have created certain measures for eradicating the invasive species, as well as preventing them from spreading further to unaffected areas. Below, we will discuss some of these species.

Figure 9.6: Invasive species are well known for their notorious damaging activities. Many of them are also competing for food sources and end up driving

away native insects from their habitats, and this can be a major reason for concern. These species not only drive away native insect species but also damage food sources for both other insects and humans. They mainly get introduced through the transportation of infected material.

Source: https://entomologytoday.org/2018/06/21/invasive-insects-the-top-4-most-wanted-list/

Aedes albopictus: This is also referred to as the Asian tiger or forest mosquito. This insect species originated from South Asia, and it has now spread to different parts globally mainly due to travel and transportation. This mosquito can lead to the spread of certain diseases such as Chikungunya fever, dengue fever, and yellow fever, together with being a host for the Zika virus. It has black and white stripes having a length of 3 to 10 millimeters. This insect has proven to be a threat because it ends up competing with other different insect species that breed in the same area, and this may lead to the eradication of the other important insect species.

Anopheles quadrimacatus: In North America, this is the common malaria mosquito, and it is also an invasive species. Despite the fact that it is believed not to be a malaria vector, it can lead to the reintroduction of malaria in locations where it had previously been eliminated. Regardless of the fact that mosquitos that cause malaria prefer tropical climates, it is also possible for them to survive in cooler areas.

Anoplolepis gracilipes: This is also called the yellow crazy ant, and it is an invasive species found in Christmas Island, Hawaii, Seychelles, and Australia, and this insect has led to major ecological damage. They tend to eradicate both animals and plants from their natural habitat. This insect has an aggressive behavior and high adaptability, making it a highly invasive species because it tends to be dominant in a habitat.

Anoplophora glabripennis: This is also called the Asian long-horned beetle, and it ended up in the United States by mistake due to materials used for packaging wood. It has ended up spreading to some parts of Europe and Canada. These beetles infest elm trees, willow, poplar and maple trees, damaging them. It originates from Southeastern Asia. Measures that have been set to eradicate these beetles have succeeded in places such as Boston, Chicago, Staten Island and Manhattan.

Bemisia tabaci: This is also called the silver leaf whitefly. It originated from India, but it can survive in both subtropical and tropical areas of the world. This species adapts to many different plants, making it very dangerous as it damages plants such as cotton, eggplant, poinsettia, cucumber, squash, beans, okra, and tomatoes. In addition to feeding, they also carry plant pathogens such as cassava mosaic virus, tomato yellow curl leaf virus, and cassava brown streak virus.

Cinara cupressi: This is an invasive species found in Africa and Europe. It is also referred to as the cypress aphid. This insect sucks the sap from conifers and this causes them to dry up. Cypress trees are the most affected, however, Widdringonia, Junipers, And Thuja. It is possible to control these aphids with the help of chemicals. These insects reproduce and increase during warmer seasons. In Africa, there are no natural enemies for these aphids and therefore they grow and spread rapidly.

Coptotermes formosanus: This is also referred to as the Formosan subterranean termite and it originates from Southern China, Formosa, and Japan. The transportation of wood is the main reason as to why this termite is now found all over the world. This insect is highly invasive, existing in large colonies and it also leads to major wood damage. These termites are quite hard to get rid of after they colonize certain areas, thus being a huge threat.

Linepithema humile: This is also referred to as the Argentine ant, and it originates from Brazil, Bolivia, Paraguay, Uruguay, and Argentina. In other parts of the world, it leads to the displacement of other native insects (ants), and this may lead to the disruption of the ecosystem due to seed dispersal limitation. When they eliminate other ants from their native locations, they also affect the food chain for the other animals that normally use these ants as a main source of food. This ant can also form a symbiotic relationship with various pests, and can thus be a major agricultural threat.

Lymantria dispar: This is also referred to as the gypsy moth, and it has many different subspecies in different parts of the world. This insect destroys trees in different forests through its notorious defoliation activities. More than 600 species of shrubs and trees are infested by these moths, and this makes it even harder to control them. They also lack natural predators. When the trees are repeatedly defoliated, they become dry, and this leads to the creation of an ecosystem imbalance.

Pheidole megacephala: This is also referred to as the big- headed ant, and it originates from many different countries in Africa, and from here it has spread to other continents through transportation. North America (mostly Florida), and Australia are suffered the most from this invasion. This ant has a disproportional body, where the body is much smaller than the head. In areas that it invades, it ends up eliminating other native ants from the area.

Solenopsi sinvicta: This is also called the red fire ant, and originates from South America. There was an accidental introduction of this ant to the Caribbean, some areas in Asia, New Zealand, Australia and North America. These ants have the ability to dwell in many different habitats, such as disturbed areas, electrical instruments, forests and even deserts. In some people, a bite from this ant can trigger anaphylaxis, thus posing a bigger threat to human beings. If not treated, this disease is fatal.

Trogoderma granarium: This is also referred to as the cabinet beetle, and it is originally from South Asia. It has become highly destructive and invasive for dry seeds and foods. This beetle has the ability to fully adapt to dry conditions, and it can also survive for long periods without feeding. It also has resistance to many different pesticides. When introduced to a new area, it causes extensive damage and may end up leading to gastrointestinal and skin irritation when consumed. To make sure that it does not invade further, food from countries that are infested is thoroughly screened before imported.

Vespula vulgaris: This is the common wasp, and it originates from the Palearctic region, however, it can easily adapt to various different habitats. It has proven to be highly invasive in Australia and New Zealand. It normally competes with other insects for nectar, especially in New Zealand. It is able to pose a high level of aggression in competition with other insects.

Figure 9.7: Research has shown that the *Vespula vulgaris* uses odor to find and attack other wasps from rival colonies, and there is a frequent change in nest

odor. These wasps have been found to aggressively compete with honeybees for the honeydew that is secreted in New Zealand's South Island black beech forests by the scale insect Ultracoelostomabrittini.

Source: https://en.wikipedia.org/wiki/Vespula_vulgaris

Wasmannia auropunctata: This is also referred to as the electric ant, and it is a tiny golden-brown ant originating from South and Central America. This ant has invaded many other countries including the Pacific Islands, Australia, North America, Cuba, Puerto Rico, Gabon, and Cameroon. In certain places that have been invaded, it has led to a reduction in insect species diversity, eliminating certain insect species, for example, arachnids.

9.5 POLLUTION

Many of the world's major cities contribute to a high level of pollution that has now proven to be detrimental to insects and plants. Studies have shown that when plants have high exposure to nitrogen oxide (NO_2) – such as the levels that are seen in these major cities – they become much more capable of defending themselves from herbivorous insects. Scientists have been able to prove that plants with higher exposure to raised levels of pollution tend to have the production of many more chemicals of defense in their leaves.

When insects choose these leaves to feed on, they tend to grow in a poor manner, suggesting that the increased levels of pollution may be leading to the causation of negative effects on herbivorous insect communities. When it comes to humans, Nitrogen dioxide has also proven to be very harmful to our health.

Insects have proven to be highly important to us in terms of nature and ecology, in this world we live in. They are very important in ensuring that our ecosystems function in a healthy manner. There may be a high level of awareness on the fact that insect pollinators, for example, many different species of butterflies, moths, flies, and the most important of all, bees, are very important for the production of food – however, they are also important in ensuring that wildflowers, trees and shrubs survive for longer time periods.

Herbivorous insects (insects that feed on plants), help in returning nutrients in plants to the soil, and they have also proven to be food for mammals, reptiles, wild birds and also other insects. They have also been proven to be very important in the decomposition of organic matter, thus ensuring that the soil remains healthy.

A major component of smog is nitrogen oxide, and this is just but an example of the pollution that we as human beings cause through our daily activities, and it is mainly due to our constant use and dependence on fossil fuels. In major cities, the levels of this pollutant are still quite high in the atmosphere, more particularly in the United Kingdom. Pollution is highly dangerous to our environment, and it is a problem we should aim at tackling if we are to survive.

There has also been some research that was based on whether insects are able to affect the ability of plants to take in nitrogen oxide from the atmosphere. When insects feed on plants, these plants tend to absorb less nitrogen oxide. This may be proof that insects have an influence on the level of pollution that is eliminated from the atmosphere by urban green spaces.

Gaseous pollutants such as nitrogen oxide can be absorbed by urban trees, however, the effect varied in different locations and with different species, and this may have been caused by the action of insects that feed on leaves. There is emphasis however on that the major concern would end up being for the insects themselves, and more research on that is still being done. Some of the research that has been done is suggesting that nitrogen oxide is take up in high numbers by urban vegetation, and many different factors can influence the effect that they have on the quality of air, inclusive of herbivory. Insects that feed on plants, however, are facing many different human threats, air pollution just being one of them. Interactions between herbivores and the direct uptake of atmospheric reactive nitrogen are mediated by plant defenses.

Another of the overlooked by a huge driver of insect population decline is light pollution. Research has shown that light pollution is leading to the death of many different important insect species. All aspects of an insect's life can be affected by the use of artificial lights at night. These lights can lure moths to their death as they fly around bulbs, as well as providing toads and rats with the opportunity to see and catch they prey faster and in a much easier manner, and even in observing the firefly mating signals.

It is true to say that Insect Decline is being driven by artificial lights at night, in combination with climate change, invasive species, chemical pollution, and loss of habitats. The worst part is that artificial lights at night is overlooked most of the time, yet it could be the main cause for an insect apocalypse.

However, in comparison to the different Decline drivers, this form of pollution is easier to deal with, especially through the simple steps of switching off lights that are not necessary, and the use of proper shades. When we do this, the loss of insects could greatly reduce in an immediate manner. Some of these lights include gas flares from the extraction of oil, and streetlights, all betting lights that are caused by humans. These lights can negatively affect insects in all the different stages of the life cycle.

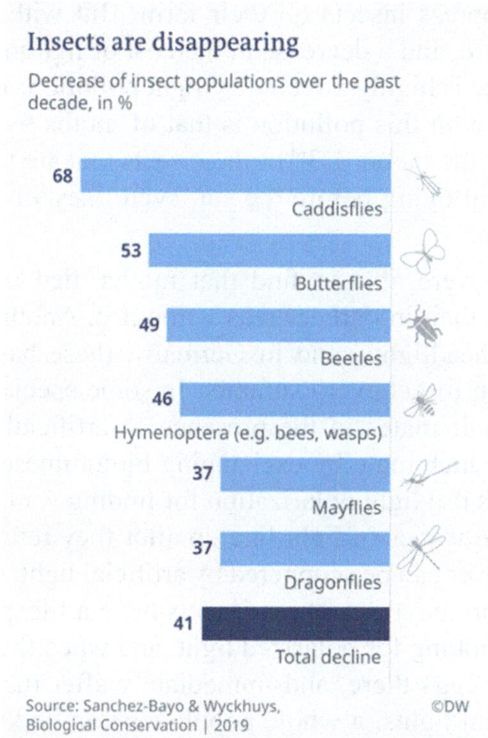

Figure 9.8: Above shows some of the insects that are disappearing at a high rate. Pollution is accounting for the majority of this decline, mainly due to global warming which is either driving the insects away, or reducing their life span. By 2019, the total decline was 41% as seen above. Caddisflies had shown the greatest decline.

Source: https://amp.dw.com/en/how-to-stop-an-insect-apocalypse/a-47723711

Puerto Rico and Germany have recorded some of the highest numbers when it comes to insect population Decline, and this can lead to the collapsing of nature's ecosystem. Insects all over the world are declining in a rapid manner, and this planet would suffer major consequences if they went extinct. Many different insect species have not yet been discovered on this planet, and according to researchers, half of them are nocturnal. Those that perform most of their activities during the day may also undergo disturbances at night as they are resting.

It has been seen in different studies that farmers have used lights for many years to suppress insects on their farm. But with the expansion of human infrastructure, and a decrease in the cost of lighting, a quarter of the world's land surface is highly affected by light pollution. The most common phenomenon seen with this pollution is that of moths flying around bulbs, mistaking them for the moon. 1/3rd of the insects that are trapped during this phenomenon end up dying before the sun even rises, either through being eaten or exhaustion.

Entomologists were able to find that moths died more in areas that had light pollution than in darker areas without it. Another deadly moving hazard is vehicle headlights, and in Germany, these have resulted in the death of 102 billion Insect every summer. In some species, insects are also not able to find their mates in the presence of artificial lights, especially in firefly beetles, which mate by exchanging bioluminescent signal as they court. Other insects use light polarization for finding water that they require for breeding, as the waves of light like up after they reflect from a smooth surface. This however can be scuppered by artificial light. One of the saddest cases can be seen in mayflies. These insects have a lifespan of a day. They mostly come out looking for polarized light, and when they find it, they end up laying all their eggs there, and immediately after that, they die. In the presence of artificial lights, a whole population can be knocked out in just 24 h.

Juvenile Insect development, for example, the development of field crickets, has also been seen to face threats from light pollution, changing the perceived length of night and day. This means that the time these insects search for food is affected. Many insects species avoid the light when foraging for food, for example in New Zealand, the giant flightless cricket (weta) end up spending very little time looking for food in areas that have light pollution.

For many species, insects are important outer, and light pollution can end up tipping the balance in the favor if the predatory species, seeing as light makes it easier for the insects to be seen and trapped. Cane toads, geckos, shorebirds, rats, bats, and spiders have all been known to feed around areas that have light pollution, and this increase in predatory risk can lead to even a faster decline in the insect species drawn to light, many of them which are important for the proper balance if the ecosystem.

Research has shown that insects have a hard time dealing with light pollution. The majority of the threats were as humans cause have natural analogs, for example, invasive species and climate change, meaning that it is possible for some insects to adapt. However, the daily cycle of dark and light has been consistent for all of the time. Despite the fact that insects may not be able to deal with this form of pollution, as humans, it is the easiest for us to deal with, cause all it requires is the turning off of a light. There is no need to clean anything up, as seen with many different pollutants. This does not at all mean that artificial lights should be banned, they should just be used in an appropriate manner.

This involves turning off lights that are not being used, and also the creation of lights that are activated by motion so as to reduce the level of light pollution. Installation of proper shades on lights so that there is the only illumination of the needed area is also important, in addition to the avoidance of blue-white lights, which highly interfere with the insect's daily rhythms. How can LED lights that be tuned to prevent flicker rates and colors that are harmful to the insects? There is strong evidence showing that light pollution has major negative effects on insect populations and thus on the ecosystem. The society is required to take the necessary steps in making sure that insects have much safer environments, to promote insect conservation.

The most appropriate step would be a national light reduction target that is enforced by the law. The insect species, population and even at the community level could have major ramifications due to light pollution. Scientists need to post more attention to this matter. It may not be feasible however to say that this impact is much more significant than the other causes of insect population decline.

9.6 PESTICIDES

Many different pesticides, such as herbicides, fungicides and insecticides, can easily be purchased at supermarkets, DIY stores and even garden centers. Untrained members in the public end up buying and using them without even taking time to read and understand the instructions that they come with, meaning that they may end up overusing them, or they may also end up using them without wearing personal protective equipment such as gloves.

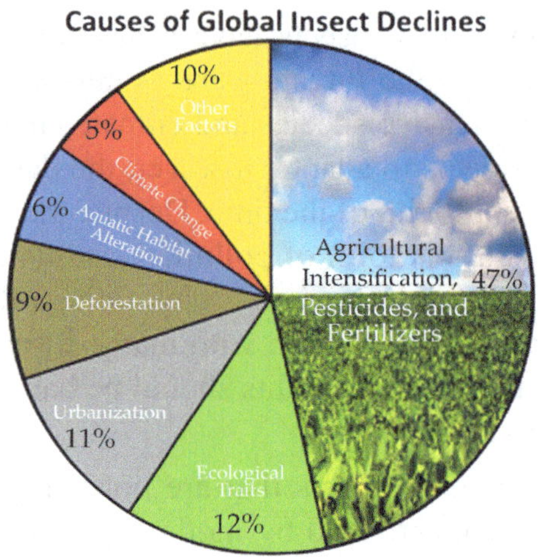

Figure 9.9: As you can see above, the use of pesticides, fertilizers and also agricultural intensification accounts for the major threat to insect populations, being the biggest challenge in the conservation of insects.

Source: https://hygeia-analytics.com/2019/02/13/intensive-agriculture-is-the-root-cause-of-worldwide-declines-in-insect-populations-new-global-analysis-shows/

In intensive farming, the pesticides that are used have shown to ay a huge role in causing the decline of many different insect species; especially because the main purpose of these insects is to kill. The use of pesticides in the United Kingdom has been documented to be much better than in other parts of the world. Despite the fact that the tonnage of the use of pesticides has declined in the past 26 years, the more disturbing trend has proven to be the increase in the toxicity of the used products, the amount of treated land, and the number of the application.

Some insecticide used, especially broad-spectrum insecticides such as pyrethroids, organophosphates and carbamates can lead to a decline in the population of important beneficial insect species such as beetles, spiders, bees and butterflies. The majority of these insects species have a huge role to play when it comes to being natural enemies of pest insects, as well as in the food web. There has been a major drop in insect numbers since 1970, especially in cereal fields (almost by half). The numbers of beetles, spiders and bugs were significantly higher in fields that had not been tested. Species and number richness of butterflies have been seen to be higher in organic farms than in conventional farms. The number of spiders and carabid beetles was also much higher in organic farms. Natural enemies are highly affected by conventional management practices in comparison to other insects or target pests. There is a higher abundance of moths on organic farms in addition to having richness in the moth species. The use of insecticides was very important in arable lands, influencing epigeic spider communities. In farms that have increased use of pesticides, communities of spiders, wild bees and bugs were much more uniform, showing that there is a reduced exchange between communities in farms that have intensive agriculture.

According to statistics that were provided in 2015 by the Department for Environment, Food and Rural Affairs, there were applications of pesticides of approximately 17.5 per hectare of land that is arable, which means that a total of approximately 17 thousand tons of active ingredients were used.

The numbers that have been provided above are not inclusive of the pesticides that are used for domestic use, along road verges, streets and parks. There is an approximately 18,000 tons of poisons that are poured on land every year, it is actually hard to believe that our animals are not already extinct. When pesticides are used, they end up accumulating in the soil, seep through and leak into watercourses, and may even end up drifting into hedges. Recent research has also proven that even the smallest amounts of pesticides used can end up causing impacts that are highly complex and unpredictable on the behavior of insects.

Majority of the insecticides used globally are neurotoxins, and these can end up leaving the bees that didn't die confused and dazed preventing them from getting back to their hives. In addition to this, there is a high possibility for interactions between different pesticides used together, and even between these pesticides and certain stressors, for example, diseases, and in the presence of two or more stressors, their combined effect may end up being more lethal and increasing the negative effect altogether. For example, the immune systems of bees may be impaired with the use of a tiny amount of neonicotinoids, making them susceptible to certain diseases such as the deformed wing virus. This information is not captured adequately nor accurately by the pesticide regulatory process, which instead ends up focusing on the short term exposure of organisms that are otherwise healthy of single chemicals.

Some of the commonly used pesticides that have proven to be very dangerous to bumblebees, bees and other beneficial insects include neonicotinoids (clothianidin, thiamethoxam, imidacloprid), pyrethroids (eg. Cyhalothrin, cyfluthrin), organophosphates (eg. Fenitrothion, dimethoate, diazinon, chlorpyrifos), and carbamates (eg. Methiocarb, carbofuran, benomyl, and aldicarb. Recently, bee poisoning has been caused by clothianidin used in the treatment of seeds. Bee behavior can also be altered by Imidacloprid residues found in plants.

10

Policies for Insect Conservation

CONTENTS

10.1 INTRODUCTION

A policy by definition is a system of principles that have been thoughtfully produced through weighing options to guide decisions and achieve a logical set of outcomes. It is a statement of deliberation and Intent implemented through procedures protocols and other tools. Policies are usually developed and then adopted with a governing body within an organization or a state and they exist both subjective and objective decision making. Promise objective standpoint policies can help senior management or leaders in general to make decisions based on the myriad of factors and the relative merits of everyone involved while on the objective level policies are made to reach an operational nature that is functional. They do differ from rules or laws however they can compel or discourage behavior and guide actions towards a certain desired outcome.

Policy studies can be studied as a subset of political sciences where we analyze the processes of policymaking the formation of a legislator and the building of institutional guidelines but in this specific chapter we will focus on policies as it pertains to insect conservation which we will find is conservation in general (Lugo, 2006).

In relation to conservation, we know that environmentalists have continuously urged the public to think on global skills but act locally meaning the philosophy behind this particular call is considering the health of the planet overall but taking action on your own ground or in your own community. As rational as this may sound it has been an approach that is delicate to balance especially with a globally connected world with quite a number of unintended consequences. An example would be with Fisheries ecology and management were in the US there were policies built two words curtailing the catching of swordfish even through bycatch. Bycatch is when fishermen accidentally catch the wrong fish or the non-targeted species that are in the ocean like sharks sea turtles and dolphins. When the Pacific swordfish was protected through the policy the US wholesalers adjusted by importing swordfish from other countries and became the largest importer of them. These and a lot more and some of the unintended further consequences of policies that are not well thought of and it can go so far as to actually transferring environmental harm where people often assume that the problem has actually been solved but instead there's a shift in impact. Another example would be how the Western world the US, for example, indicate timber harvest reductions as much as even 30 to 40% however when properly observed it has only been shifted to less developed countries including tropical forests in South and Central America

When it comes to policies on the conservation of biodiversity the basic policies provide fundamental rules that lead us towards the implementation of activities with considerations paid to the said biodiversity. Kao operation, for example, is a chemical and cosmetic company based in Japan with a global reach and they have their own basic policies on the conservation of biodiversity which include identifying the relationship between their business activities and their environment so as to reduce the impact or at least the negative ones on the environment around them. This involves complying with the relevant international agreements that already exist but further includes a public relations outreach on pursuing local restaurants within the vicinity of the businesses as well. Many companies and corporations have something similar along these lines and how they themselves interact with the community around them and the ecosystem and biodiversity (Lugo, 2006).

On the country level you will have policies acting in the same manner but probably on a higher scale because they have a part to play when it comes to the formation of legislative and institutional tools that go towards conservation and restoration of biodiversity period an example would be the government of Trinidad and Tobago that recognize the needs for effective implementation of policies to monitor regulate and manage the environment ecosystems and more especially in relation to government services in order to reverse environmental degradation which ended up forming their 1998 environmental policies and regulations which was built up over a hundred different laws that were deliberately thought out of in response to environmental problems that they faced. The policies in question that were developed in relation to enhancing sustainability management of the National ecosystem with resources with the first one being formed in 2006 named the national environmental policy.

Writing a policy includes developing a set structure with a list of different components that include:

- A statement
- Principles values and philosophies
- A broad objective
- Strategies to achieve the objective
- Actions to be taken
- Desired outcome from the actions
- Performance indicators management plans

- A review program

Policies can be very brief but also some are complex in nature with the important thing to realize in the formation is that there is no substitute for research and consultation when developing what would be an effective and successful policy.

So with this in mind, the Trinidad and Tobago national environmental policy had a broad overall statement that was listed as their goal which was to ensure environmentally sustainable development in order to enhance the quality of life and meet the needs of both the present and future generations. Their objective through this policy was to conserve the vitality and diversity of the national environment to conserving ecological systems and the various biodiversity that lies within them. The specific strategies in this policy to meet their objectives and their goal included several tools like the establishment of a system for protected areas and a system that coincided with it for the Protection of Animals and plans to the designation of environmentally sensitive areas and environmentally sensitive species respectively. Another strategy that they had to meet the national environmental policy of 2006 was the conservation of representative examples of all coastal and marine ecosystems including them in a system for protected areas as well. They had an extensive list of specific references and tools with actions that needed to be carried out and many performance reference indicators over a period of time. The above serves as a decent example of how policies work and more specifically how an environmental policy would be formulated (Lugo, 2006).

10.2 INSECT CONSERVATION

As most probably stated in previous chapters, insects have been an immensely successful group of animals who dominate the compositional diversity that exists on the planet Earth probably spread across the globe in all areas including those that are extremely salty or extremely cold. There has however been over the recent years what is now commonly being referred to as the great insect dying with many colleges are environmentalists calling it the sixth mass Extinction event to happen on planet Earth. Insects make up nearly 70% of all ,for now, that exists on Earth but are dying in a rate that is very alarming due to the various human activities that are happening today and over the last several decades or even centuries

The precipitous decline of insects on Earth should be a great concern ecologically as insects contribute to almost all ecological systems that exist on earth from the maintenance of terrestrial and freshwater systems to the service delivery of pollination and other functions. from a philosophical perspective, we can argue that Extinction is the norm and 99% of all organisms on Earth that have gone extinct have done so by natural causes so it can be argued that events should be left to take the course for the evolutionary purposes of the planets however, there are certain things to consider. First off is the intrinsic value which we have already started to some extent although quite briefly. We as human beings should be wise and caring two words all life that exists on the planet. Secondly, it should also be noted yet again that insects play an instrumental value in all functions of the environment even though so few people actually appreciate some actually looking at insects with disgust and disdain (Collins, 2012).

When relating to insect conservation and policy-making, there are two main ways to consider the direction. The first would be to look at insects as a component of biodiversity and support their survival with the second being in that insects provide essential ecosystem services and we cannot survive without them. Both these perspectives can be looked at and considered on a variety of spatial scales in exactly what we meant when mentioning thinking local and acting global. This must be however put a cross -bearing in mind that the nation-state themselves for their own sovereignty and powers and so they are the most important in terms of conservation action unit.

The species survival commission under the international union for the conservation of nature as an example of insect conservation policies on a global scale

When looking at the formation of policies for in the conversation from a biodiversity perspective insect conservation is integral in biodiversity conservation because they have so many functions at so many different trophic levels and interact with a variety of other organisms from fauna like pollen- producing plants fungi and even bacteria and microorganism like protocysts and this includes within and among themselves. Biodiversity conservation is no new phenomenon and is coined on a global level in terms of the convention on biological diversity (Collins, 2012).

Insect conservation policy has various perspectives. Overarching is the Convention on Biological Diversity to which most countries in the world are signatories. The Aichi Biodiversity Targets, which are specific targets, mostly have great application to insect conservation. In terms of conservation of species specifically, you have the International Union for the Conservation of Nature's global red list gives important coverage to insects that are threatened. At the national level, local red lists (RLs) are important too. National Biodiversity Strategy Action Plans (NBSAPs) are also significant for local insect conservation, especially in biodiversity hotspots. NBSAPs cover a wide range of biodiversity issues and targets besides insect species conservation.

The convention on biological diversity also known as the biodiversity convention is a multilateral Treaty agreed on by 96 different states with ratification by 30 States and it was a convention held in Rio de Janeiro Brazil and New York the United States as well. It's offered some specific goals but excluded summer shoes especially pertaining to Marine environments and insects' conservation. On a global scale, insects are conserved at the species level and in terms of insect diversity. At the species level the international union for the conservation of nature and the species survival commission are both major proponents on the wider scale by setting out standards and actually producing the red list which is a list of threatened species that exist that is easily accessible online. This globally-recognized list provides an inventory of the Worlds species that are currently under threat and furthermore studies by researchers are done and suggestions are provided on any conservation action that can be taken. The way this is done is through activities and a network of scientific specialists under various taxonomic groups (Collins, 2012).

This is a function of policies and this is how they work there is the overall objective or goal that is set after recognizing a problem way sitting bodies Like the aforementioned species survival commission who play their part in researching and providing further information on how to reach the policy objective with constant review or performance indicators which in this case is served as the red list of threatened species. The red list serves as the direction towards action to be taken however it is not a priority list as such and nor is it a political document that stipulates or gives instructions to what governments or agencies should do towards their own conservation efforts of insects however and probably even more importantly it is a general assessment of the status of conservation as it is happening for the insects and various other animals that have been assessed. It is important and serves as a highly influential policy -making body towards the conservation of insects on a global level. It serves well for insects in this case as they would receive the same treatment as let's say wombat or a whale if their profile is raised and by the extent they will find the way into policy documents that are being developed for biodiversity conservation.

One of the challenges here for insect species that are raised in the red listings is that it is intentionally specious. Today about 7700 species have been evaluated for the red list which is about 1% of the described species that exist on Earth of about 1 million in total. Those that do make it into the red list are probably less than 0.2% of the millions that exist. The reason for these low numbers is that the extent of fieldwork that needs to be done is so considerable and to assess these three statuses of insect species requires specialists and there are relatively few that exists today to do that. The situation is aggravated by the fact that many of the Species go extinct without even ever having received a scientific description or scientific names and this is a phenomenon that is known as the Centennial Extinction. This event is named after the Centinela ridge in Ecuador where many botanists and ecologists found plants that went extinct from deforestation before they were ever scientifically described (Colla, 2017).

The insects that have been analyzed and assessed need to be quantified so that we can get a sense of the various kinds of threats that they may be facing. When grouped with terrestrial invertebrates, in general, the main threats that insect face today are the following in order of significance

- Habitat loss due to logging and other related activities
- Habitat loss due to agriculture
- Habitat loss as a result of infrastructure development like urbanization for example
- Habitat loss and fragmentation due to the construction of service corridors and transportation
- Invasive alien species
- Change in fire regime
- Pollution
- Climate change and adverse or severe weather effects
- Mining

As state above this is more for terrestrial invertebrates however for freshwater invertebrate's pollution and poor management of dams for example are the major threats that they face somehow even greater than all the others listed. It is important to know what threats that they face not only for future planning in conservation but also to realize that for any species of insects there exists more than just a single daunting threat that they face and these threats are often synergistic in nature.

The convention on international trade in endangered species is yet another effect that came through the international union for the conservation of nature drafted as a result of it and adopted through a resolution in 1963. The convention on international tradc in cndangcred species was opened for signature in 1973 enforced in 1975 and is yet another body formed through the directives of the United Nations and it is a multilateral treatment that is directed towards farming regulations in the trading of endangered species of wild fauna and Flora and is simply called the Washington convention. Through various records with different degrees of protection about 35000 species of animals and plants have been protected especially having been threatened with extinction and to ensure the policies that are piled the general agreement on tariffs and trade were reviewed and consulted through the drafting process of this convention. In regards to insect store, very few species are listed and they include mostly quite large, conspicuous and highly collectible or charismatic species like the queen Alexandria's birdwing butterfly, the largest butterfly on earth (Colla, 2017).

Figure 10.1: *Ornithoptera meridionalis*, the southern tailed birdwing, is the smallest species of the genus Ornithoptera. It is known from a handful of localities in southeast Papua, New Guinea and several localities along the south coast of Irian Jaya. Its bigger counterpart, Queen Alexandra's Birdwing Butterfly is also protected in under the Convention on International Trade in Endangered Species (CITES).

Source: https://www.activewild.com/queen-alexandras-birdwing-butterfly-facts/

10.3 INSECT CONSERVATION POLICIES ON NATIONAL LEVEL WITH GERMANY AS AN EXAMPLE

So far we have been discussing insect conservation and their policies in relation to the global scale but when you move down to the National scale, insect species conservation is greatly varied with different countries and you will find countries that are more financially stable with a higher gross national product will more often than not Focus their attention and Resources to insect species conservation than those that are more financially constrained with obvious reasons. Some countries have well-placed and enforce regulations that protect insect-like Germany and their protection of dragonflies. The environmental Minister was very quick to note that we humans need insects and they deserve protection from their own law and the insect conservation policies need to extend beyond the protection of stag beetles and earth bumblebees and with the Insect Protection Act regulations were actually enforced changing the rules governing nature conservation water resources and plant resources continuing to further restrict the use of artificial fertilizers and developing laws that form part of the environment Ministries insect protection action plan with 25 million euros a year being founded towards insect research (Colla, 2017).

This is a great example of the practicality of how governments or more local implemented policies would translate into regulations and legislature with adequate funding towards the protection and conservation of insects. It is important to note that although countries like Germany, the US and the UK are Pioneers in the conservation of insects in production of policies they work in hand with international bodies especially, the United Nations in the formation and pose as an example in the implementation of these policies especially after they have been successful in their home states. In Europe for example this idea of the insect Protection Act had a further impact on various consortia that are in the European Union most notably the European Union habitat directive where there has been developed cooperation on the protection of insects across the European continent.

This goes further with the Millennium ecosystem assessment that was developed by Kofi Annan when he was a Secretary-General of the United Nations where insects were classified as being highly significant and where provided with provisional services noting insect importance in biological control of pests as well as ecosystem monitors and even in the tourism industry in attracting Revenue. Policies in insect conservation are often formulated in realizing the importance of insects as pollinator's seed dispersal contributors of atmospheric gases and providing support services in breaking down dead plant material and animal material into the soil. This provides justification for the conservation of insects and their full justification for the formation of policies and funding two words efforts made in conserving them.

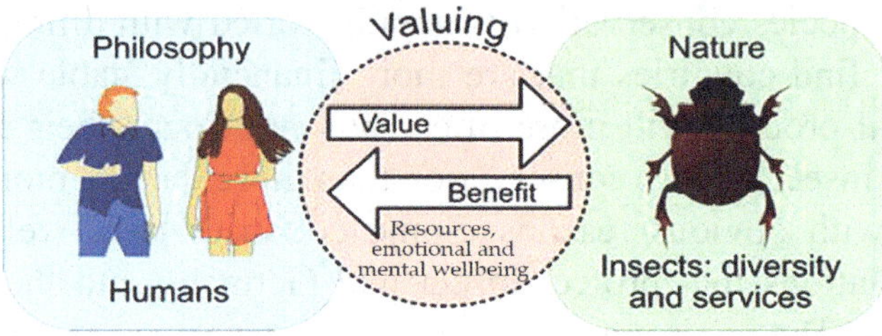

Figure 10.2: Justifying insect conservation and acting on this justification requires a philosophical view based on value systems on nature. Valuing the human relationship with nature results in benefits accruing in terms of physical resources such as food and also in terms of well-being on the mental, emotional and even spiritual levels.

Source: https://www.intechopen.com/books/insect-science-diversity-conservation-and-nutrition/insect-conservation-for-the-twenty-first-century

On a positive note, it is encouraging to note that the recognition on the value and importance of insect has been forwarded into insects being enshrined more and moreover the years into policies that are being formulated in general the conservation of the ecology and biodiversity and in some countries especially those that have a deep respect for nature they recognize this more so. It is a system that has been complex in its development especially when you notice that entomophagy which is the consuming of insects by humans has existed for time immemorial. Today an estimated two million people eat insects as part of the traditional diet and have proven to be highly nutritious and high in protein and fatty acids when compared in weight with the conventional beef or chicken counterparts. Moth larva, beetles, bees, wasps and ants are some of the more commonly eaten insects becoming increasingly more important in Direct human consumption and for livestock over the last several decades and have been used as a way of tackling malnutrition in children as well successfully, to say the least. The red meat production is resource expensive in comparison to insect production as food and even more so as global populations rise and resources become scarcer. In the past, they have been traditionally harvested from nature in the wild but initiatives have been made towards rearing them on large scales. this has its own challenges, to say the least however new techniques and processes are being developed and somehow it has its own past to play in the insect conservation and protection perspective because it would be remiss to obviously it's an insect that is under threat and those that would be consumed would be those that naturally occur in abundance. Insect farming and its improvements will also go hand-in-hand with developments and appearance visually of them as food to make them more acceptable to the human population generally (Kothamasi & Kiers, 2009).

10.4 INSECT CONSERVATION POLICIES AND THE PSYCHOLOGY

As previously stated environmental policies on insect conservation rely on a dichotomy of two mutually exclusive value systems which are the instrumental valuation and the intrinsic valuation. This is helpful to a certain extent as simply having an overall value of nature and appreciating both extents will help developer platforms or appropriate policy meetings as much as the human relationship with nature is very complex. The instrument valuation approach system is more quantifiable especially financially and this can create more overt drive. You will find however with the intrinsic value systems especially with people having been affected by climate-related impacts there is motivation towards insect conservation. This should however take into account that politicians and policymakers have to be accountable and making that relation between anthropogenic factors natural factors and how they have subsequently affected the insect population on Earth. The importance of Raising insect conservation policies psychology is to understand and policy-making by conservation bodies either national or International in its formation stages when it comes to making a statement deriving principles values and philosophies outlining a broad objective and formulating strategies to achieve the objective.

One example we can take a look at is the Xerxes Society for invertebrate conservation which is an international non-profit organization that protects the general natural world through conservation of invertebrates and more so their habitats. It is a science- based organization. The principles and values are three-pronged in their approach when it comes to protecting the Natural World to the conservation of invertebrates and their habitats and on their website, these are listed as the pollinator conservation program that works with partners restore and expand habitats of pollinators the endangered species conservation that is more engaged words raising awareness and education researching on various invertebrate species carrying out community science and advocacy to protect at-risk species and their habitats and the third pillar is reduced in pesticides and its impacts on the insect world by educating agriculturalists on reducing their over-reliance as it has been over the last few decades on pesticides by supporting a more diverse natural system and engaging in biodiversity that would work in reducing pests as a problem. This is the philosophy of the Xerxes Society and we are one of the many organizations that carry out activities through funding by the government and other foundations so far receiving 129 grants that in total are equivalent to 4 million dollars (Colla, 2017).

Biodiverse is a European best conservation organization that focuses on the conservation of threatened insects in Europe managing their habitats and climate change adaptation. It is also based on a group of scientists that carry out research projects on biodiversity and provide recommendations to policymakers based on the results. On a positive note, they have recorded that the application of conservation measures has made it possible for the large blue which is a species of butterfly in the family Lycaenidae to become more popular in the United Kingdom also to the conjunction of other scientific products prior to it.

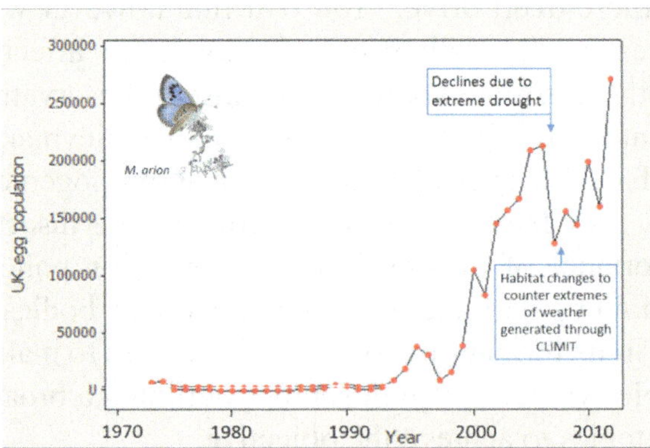

Figure 10.3: The abundance of the Large Blue butterfly in the United Kingdom, preceding extinction in 1978 and following its reintroduction from Sweden to carefully managed sites in the 1980s. No precise data are available for numbers prior to 1972, but anecdotal descriptions indicate a continual decline, with a similar number of butterflies to today having last occurred in the early 1950s.

Source: https://www.iucn.org/downloads/biodiversa_pb_2_conservation_of_threatened_insects_in_europe_web_1.pdf

In relation to policymaking research cannot be overstated it is as important as the policy itself and that is why performance indicators and management plans as well as constant reviews are part of that scientific process as well as its practical political implementation (Colla, 2017).

Some of the strategies used by non-profit organizations to protect and conserve insect biodiversity include working with various governmental bodies to incorporate the needs of pollinators for example and other invertebrates international conservation programs non-profit organizations need to work together with lawmakers to pass legislation to improve habitats for invertebrates and insect species and promote their protection using existing laws such as the endangered species act in the United States of America and some of the recent work that has been done along those lines or efforts to include pollinator conservations in the farm security and rural investment act which is otherwise known as the farm bill.

Pollination by bees is critical to the production of very many crops with more than 30% of the food produced in the United States alone relies on insect pollination and the annual value of crops pollinated by Wild native bees and honey bees in the US estimated to be around 3 billion dollars and fifty billion dollars respectively. The native bees ,however, have declined another region due to Habitat loss which we have listed is really the biggest challenge and cause towards the great insects dying and at the same time colonies in Europe have also suffered a 50% decline in recent decades which subsequently leads to the Threat of invasive pests and diseases. The economic importance of pollinators has already been stated with the instrument valuation perspective and non-profit organizations together we government founded ones like the national research council are together able to partner up 2 push forward campaigns like the North American pollinator protection campaign that was able to convince legislators in the United States Congress to pass the pollinator Protection Act and the pollinator habitat protection act both of which were fitted into the farm bill legislation and passed by the House of Representatives and the Senate (Lugo, 2006).

10.5 INSECT POLLINATOR CONVERSATION POLICY INNOVATIONS IN THE US

Just like the rest of the insect group insect pollinators in specific have nobody been on the decline mainly due to human behaviors and land use and policies have been directed towards rectifying the situation as a mutually agreed-upon limit to behavior to achieve shared values are necessary for tackling this complicated social -ecological problem. In spite of the scientific calls and generally the public outcry that has been going over to develop policies to address the decline specifically multi-lateral decision agreements and conventions have not been able to deliver proper legislation or even to basically beat monitoring needs that have been recommended by experts. In this section we will hope to look at policy innovations that may be able to change the direction of this challenge it is also more specific to have a good scope to tackle these issues and this specific reason we will look at insect-pollinated conversation policy Innovations at sub-national levels especially in relation to turn in his policies into legislation rather than its setting as per your recommendation.

The reason why we have decided to focus more on insect pollinators it's because nearly 90% of the world's flowering plants depend upon insect pollination for reproduction and all animals on the planet including human beings rely on these plants as a source of sustenance. This is more of importance not only psychological even looking at the instrument value or intrinsic value because they pretty much fall into both quite heavily so and the declines due to several compounding factors are very alarming with studies done by the intergovernmental science-policy platform on biodiversity and ecosystem services showing significant evidence that over 40% of the world's invertebrate pollinators are threatened or are at risk of extinction but mostly looking at bees and butterflies like for example the discovery of 84% decline in population for the monarch butterfly just towards the end of the 20th century which is quite alarming. Although pollinators include several taxa which we can provide examples of including bats beetles moths and various different birds and not just in the insect order we do have bees especially being the most effective animal pollinators that exist on the planet responsible for some believe up to 90% of all global pollination (Lugo, 2006).

Response to this global decline and a true realization of our dependence on insect pollinators various governments across the world seems to be marshaling policies to cut off the losses and two regrow their native insect pollinators. Unfortunately, the True Match of the fact is that despite calls for pollinator policy to address the pollination crisis as well as invertebrate conservation policy advice there really hasn't been a scientific study of insect -pollinated policies that can be well -reviewed and this needs to be done in order to direct lawmakers towards the problem and evaluate the general design and adoption of these conservation policies. Will this daunting fact we can still say that the US congress has passed within the last 10 years about 4 of 31 bills presented to them that it's somewhere within them address pollinator health status in the US. The Food Conservation and Energy Act of 2008 was in response to scientific pollinator policy targets with appropriations to investigate pollinator biology immunology ecology genomics and bioinformatics that are somehow associated with the Colony collapse disorder and also furthermore they will study various other threats on honey bees are other pollinators that include parasites and pathogens that may affect them or insecticides herbicides and fungicides that would be otherwise lethal to them parents research like this should be aimed to promote the health of honey bees and native pollinators by firstly conserving their habitats and secondly having management practices that are concerned to them.

The Agricultural Act of 2014 is there another one that sort of went to words pollinator protection and actually authorized appropriation through the whole fiscal year for research in the same direction will also be fixing America's surface transportation act of 2015 encouraging pollinator habitat and development by including integrated vegetation management practices that directly helped the monarch butterflies and other pollinators like honey bees as well through the planting of natural native plants that would be stationed for them and facilitate them (Lugo, 2006).

We can also discuss the federal agricultural Reform Act that was yet another policy that was transformed into legislation and in this case, it was to improve federal coordination that addresses the decline of pollinators in general so that we can promote long-term viability. Though these laws are few in number they represent a general consensus on the problem especially with a divided nation such as the United States. But what should be noted that in the absence of a complete and comprehensive National legislation sub-national policies and assemblies become even more important? You will find that United States legislature would like to sleep anywhere in the world really are very reflective of the values opinions and desires that the populations of the Host have and consequently, policies emerge from the same ideas of the values and desires and opinions that the people hold whether they may be demographically or economically divided. Policies actually reflect economic scientific cultural and legal backdrops with sub-national laws and policies will incubate reactions of the people.

Many of these policies can be simply searched on using usa.gov and from there it is quite easy to catalog and analyze pollinator relevant policies that subsequently became new laws on a sub- national level and eventually on a US state level. When this is done you will find about 110 legal documents that fit anywhere within the searches for pollinator and policy together and essentially they were 36 state legislatures passed that effectively fit within these 110 pollinator relevant policies with 14 States not having passed anything at all which is very disappointing (Kothamasi & Kiers, 2009).

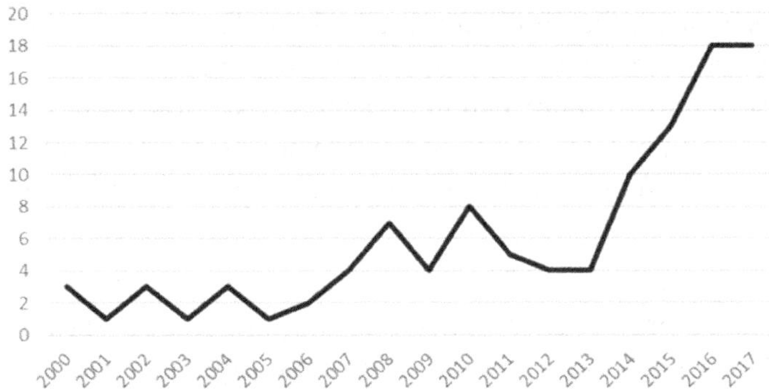

Figure 10.4: the table above shows the number of state insect pollinator policies passed and approved by Legislatures and signed into law by Governors by year.

Source: http://millionpollinatorgardens.org/wp-content/uploads/2019/02/Hall_Steiner_2019_Insect-pollinator-conservation-policy-innovations-at-sub-national-levels-Lessons-for-lawmakers.pdf

Policies in this exact relation are important in that they tighten agricultural standards on how farmers may manage diseases and pests. Secondly, they may be able to create new task forces that provide updates on the various approaches we use towards Pest Management like pesticides used for example. Thirdly they establish and improve pollinator habitat even if it is somehow in an indirect fashion and fourth the appropriations gathered will fund research into monitoring and managing all native insect pollinators. Lastly and probably most important in the long-term run is raising public awareness about pollinators in general and subsequently insect conservation policies.

State legislation actually is capable of responding to agricultural industry issues and can use policies to address various problems that may arise. On record even as far back as 1883, there has been legislation passed in the direction of the insect-pollinated preservation. This was done to address the American foulbrood which is a bacterial disease that honey bees carry Paenibacillus larva. It was included in the California State legislation for what would be now recorded in history as the first state-level apiary inspection. These authorized inspections to be carried out for the disease known as foulbrood and any apiary in the state and if the results came back positive then the person in charge would be directed to destroy the Hives by burning burning infected hives and bees. Later on, the US Congress passed a similar law known as the Honey Bee act that place restrictions on the importation of honey bees as a preventative measure on spreading diseases or parasites that are harmful to honey bees. It was later amended in 1976 to stop the undesirable species of subspecies of honeybee being produced in the States like for example the Africanized honey bees.

On a global level, the fraction of pollinator- dependent crops has increased from 3.6% in 1961 to 6.1% in 2006 as a percentage of total agricultural production and this has been realized in many countries like the US that is in discussion with quite a variety of crops depending exclusively on insect pollination like the watermelon and the cocoa plant with managed bees more specifically being effects of pollinators and therefore registration and regulation of pollination a parasites and quarantine policies have been developed almost exponentially in recent years (New, 2012).

Varroa destructor and Varroajacobsoni are two mites' infestations that have caused ailments to honey bees leading to entire hives dying off between 2005 and 2007 where will the winter it was actually estimated in the US 23% of bees suffered from colony collapse disorder with affected beekeepers losing up to 45% of the bees. From here policies were mobilized and state legislatures were formed to direct the resources towards colony collapse disorder research and management. An example that can be offered along these lines is New Hampshire in 2008 where laws were formed and appropriated funds for the purposes of inspection and additional funding was also directed two words essential research on the above-stated Colony collapse disorder with more funding also being required two words with generating the Bee population. They were not the only state and several others had similar legislation to addressing these mite infestations. In fact, it went so far as some states like Washington offering temporary business and occupational tax relief for its apiarists.

The beauty is as public awareness is raised and public calls are made towards protecting and enhancing bee populations, hobby beekeeping is becoming increasingly popular even amongst celebrities like Morgan

Freeman despite the unintended consequences that this may have to the native bee populations. Hobby Apiculturist actually presents a new angle and food concerns when it comes to policymaking and legislature in insect conservation. What has happened in that respect is that many states have had to adjust the classification of bees as livestock so that they would fall under livestock laws for the purposes of being able to tag what can be taxed. In Montana ,for example, the definition of a hobbyist beekeeper it is one who owns a total of not more than five hives and this now goes further to exempt hobbyist from actually paying registration fees for their beehives. In Hawaii, a hobbyist beekeeper doesn't actually need to have a permit for his or her hives or honey processing simply because honey bee populations are declining at such a rapid rate and by reducing these legal and financial restrictions hobbies or even commercial beekeepers are sort of incentivized (New, 2012).

In relation to insect conservation, it is obviously evident that there is an evolving view of bees and insects in general as beneficial in which true sense that they are and so policies and legislators have also gone and Revisited pesticides as well. Unfortunately, from a historical standpoint, you will find that agricultural status has discriminated insect-pollinated classifying them as crop pests or bugs and it is only recently that legislator and policies have actually been diverted to codifying law with the new notion that not all insects are pests which in a true sense many of them actually are not.

Beneficial insects are a new concept but it has become active within a policy of integrated Pest Management which fits into an overall agricultural policy related to pesticide application. The state of Maine for example as a definition for beneficial insects that is those that during their life cycle are effective pollinators of plants or parasites or Predators or pests or are otherwise generally beneficial in the Maine public act of 2006. Beneficial insects are a new concept but it has become active within a policy of integrated pest management which fits into an overall agricultural policy related to pesticide application. The state of Maine for example has a definition for beneficial insects that is "those that during the life cycle are effective pollinators of plants, are parasites or predators of pest or otherwise generally beneficial" in the ME public act of 2006. Various other legislation in different states have used similar terms and language as well and this is important in policymaking for insect conservation because it changes the overall public perspective on insects especially those that are deemed beneficial.

This has caused an evolving view of insects in general with many states over the last 20 years organizing the departments of Agriculture and Department of Environmental Management with the Environmental Protection Agency has their partner to visit the use of pesticides the application and disposal rules of said pesticides and essentially in building pesticide awareness and training. Indiana, New Mexico, main Kansas and Minnesota are just a few of the States in that list (Colla, 2017).

Neonicotinoids are a class of reactive insecticides built chemically similar to nicotine that was developed in the 80s and the 90s and his whole family of various damaging insecticides with the most famous in this family being imidacloprid. They became increasingly popular because they cause less toxicity in birds and Mammals than they do in insects. Neonicotinoid and its use over time has been linked in very many studies to have adverse ecological effects especially in the honey bee colony collapse disorder and sometimes even affecting Birds indirectly due to the insect population decline. In 2013 the European Union and several other neighboring countries restricted the use of Sam Neonicotinoids and in 2018 the three main popular ones being imidaclopridthiamethoxam and clothianidin being totally banned for all outdoor uses. In the US, there is also restricted usage and various policies that regulate the use of neonicotinoids. It is an insecticide of pesticide that is known for high efficacy against sucking insects and many chewing insects and is also lethal to some soil insects and fleas. It is mostly mixed with water and then used to irrigate plants and used for foliar treatment on animal pests as well. Sometimes it is even used on seed coatings like for example soybeans that are feasted upon by the soybean aphid although it is not effective against this particular pest.

There have been state legislations as already said above on neonicotinoids because of declining pollinator health. Although many states actually have encouraged voluntary Pest Management processes and practices that would be beneficial overall to the ecology and insect conservation other States take a more prescriptive approach like California Who banned four Neonicotinoids in 2018 due to their obvious potential hazards to honey bees. This is after studies were done showing the sub-lethal effects that they have insect pollinators which include impaired foraging and feeding behavior period insects also affected by neonicotinoids will have weakened immunity and delayed larval development which makes them increasingly susceptible to viruses, disease and parasites. The Maryland pollinator protection act of 2016 that was pushed by the Department of Agriculture forced policymakers to think of regulations that are protective of pollinators and the act entailed the restriction of retail sales of Neonicotinoids and limited application of neonicotinoids to only certain applicators and goes beyond just out to fining violators of this act (Colla, 2017).

The Kansas state department of Agriculture also recognized the damage that pesticides do showing concern for insect pollinators affected by Neonicotinoids updating regulations and strengthening the pesticide applicator license requirements that they had. Oregon also had state policies that constantly update pesticide training especially when there are more natural alternatives to take and the state of Minnesota followed suit adopting almost-identical policies. The state of Minnesota has statutes that were applied effectively of 2014 that prohibit the labeling of advertising of stop plants in Nurseries as beneficial to pollinators if they have been treated with systemic insecticides. They also updated their definition of pollinator's lethal insecticides quantifying them as insecticide that is absorbed by a plant that makes the seed plants lethal to pollinators and includes but are not limited to Neonicotinoid and similar classes of insecticides.

In 2016 to address pollinated clients, Whitehouse formed a pollinator health task force and consequently, States like Connecticut Vermont and Oregon also established a state-level pollinator committee respective to the same concerns. This also developed and evolved certain agricultural practices, Apicultural practices conservation practices and pesticide use. Oregon's task force on pollinator health was precipitated by the Wilsonville bee kill where trees were sprayed with dinotefuran that is a damaging pesticide with eventually the task force having the immediate outcome of a new department of Agriculture regulation that restricted the use of neonicotinoids.

The pollinator Advisory committee in the state of Connecticut is a little bit more comprehensive in the way that it is built and is headed by the commissioner of Agriculture who collaborates together with experts from the Connecticut agricultural experiment station and members of the Department of Energy and environmental protection where each of them hand-in-hand has a part to play to ensure the health and viability of pollinator populations in the state. They focus on developing practices that would minimize the airborne use and liberation of Neonicotinoids and generally tries to limit the amount of pesticide used by application amount (Colla, 2017).

The state of Vermont also has a similar group known as the pollinator protection committee that was initially formulated to evaluate the various causes and occurrences that perpetuated the reduction of pollinators and not just insects alone. This committee is made up of the secretary of Agriculture two major beekeepers a major dairy farmer and an entry advocate for pollinator protection. There's also a tree food farmer a vegetable farmer a licensed pesticide cell person a greenhouse operator and a university employee with pollinator protection expertise that have seats on this committee. This committee is tasked with examining other States and general international pesticide regulations in order to form their own unique regulations and policies that would protect insects and pollinators in general (Lugo, 2006).

10.6 POLICIES FOR THE CREATION, PRESERVATION AND MANAGEMENT OF HABITATS FOR INSECT CONSERVATION

Deliberate and strategic landscape planning for insect conservation to combat decline would first require one to observe where these insects live how they forage nestmate rear their young and in some cases over winter. This knowledge will arm policymakers in creating effective habitats on state land and prioritizing this for the purposes of conservation. This policy these policy actions can take several forms. State agencies can have codes of regulations that are updated especially on insect habitat priorities with the appropriation that is required. In New Mexico for example there was a memorandum that was passed that determined all state and county agencies that run from public schools to colleges and universities and any other grounds that worked using state funds to create or use existing resources for insects and more so pollinator-friendly plants and Landscaping. Certain crops grown in New Mexico like the chili and other products depend on bees and other pollinators. The native environment also has unique flora and fauna that go as far as attracting tourism bringing come to the state and could be affected by the declining insect populations and so there was an instrumental value in the formation of such policy (Lugo, 2006).

California created policies and legislation that directed The Department of fish and wildlife to conserve the famous monarch butterflies and its unique habitat they depend on. Habitat restoration was discovered to be one of the most effective and successful methods of doing this together with the use of education programs and working together with landowners two voluntary agreements. Many states have agencies that develop policies to respond in an almost opportunistic fashion using existing land management programs and slightly tweaking them to the purposes of insect conservation and Preservation.

It is important to note that policy in habitats creation and Conservation in line with working with insect preservation and Conservation extends to the enhancement of these insect habitats on private lands and this works in New Jersey ,for example, where homeowners are encouraged to plant native plants that support local insect populations and create habitats for all sorts of insects and this goes as far as the Department of Environmental Protection and agriculture working together with Nurseries and landscapers around state land to develop supportive efforts and further this policy. Minnesota has probably the most comprehensive habitat protection policy and enforcement through the insect habitat programs at directs the commissioner of Agriculture towards the development of management practices that are best for habitat restore Asian guidelines for insect habitat enhancement (Colla, 2017).

10.7 POLICIES FOR INCREASING AWARENESS OF INSECTS FOR CONSERVATION

With the creation of the National pollination week designated by the US Congress in 2006, various similar awareness directives are intended for the recognition of the emotions of insects to the ecosystem health and agriculture. The public outcry to save the bees brought to attention the same idea of insect conservation especially in maintaining the beauty diversity and enchantments of specific landscapes. By 2016 all US states actually had designated an insect-pollination week which was primarily a directive of the state executive. As much as most insect awareness policies are informative at best lacking deadlines or actionable items we also have to remember that they have a lack of appropriated funding but they do serve as a benchmark or starting point two words tailoring advanced conservation actions.

10.8 INSECT CONSERVATION POLICIES TURNED INTO PRACTICE

It is important to remember that insect conservation is a practical activity that is founded on scientifically sound research and a good knowledge base. The starting point for practice is well understood from a philosophical point of view together with the research findings and well-established strategic policies there are in response to the research that has been done not only for insect conservation but for conservation psychology in general. You will find whoever that insect conservation ,in general, is not often practiced by the philosophers, researchers, and policymakers period in conservation from a practical action point is usually perpetuated by conservation organizations farmers who are concerned and progressive companies. Responsible and stewards should also be mentioned and the ones who were cognitively aware of the philosophy behind it and importance all the policies as directed.

The importance of non-governmental organizations like the Xerces Society, Butterfly Conservation and Buglife receive some sense of appropriation or donations from private funds cannot be understated as well in the actionable channels that they take towards insects' conservation period on the global scale we have already talked about the IUCN as well who fully perpetuates the concept of thinking globally and acting local. It must be understood that these policies once they have been made r almost in a sense non-negotiable that is to say that for example proclaimed protected areas should be recognized formally and actions thereof should be legitimized in a similar fashion. The reason why this is significant is because can you find some of these landscapes or biomes been completely novel and unique and is often the last areas for many rare and highly localized endemics with a lot of very specialized interactions. When it comes to Habitat protection for insects you will find many of them are ordinarily isolated but still requiring some sort of management especially when it comes to mimicking the historic landscape as it existed. The use of Fire regimes for example is some of the management techniques that can be employed to simulate an idea of the natural conditions before human fragmentation, unfortunately, took hold of the landscape. This can consequently reduce the fuel load and risk of a potentially highly damaging fire further down the line while also stop peeing vegetation succession to an overwhelmingly Woody environment.

Some other practical approaches to creating insect conservation site are integrating protected areas with the surrounding unique setup. This could be the biosphere reserves for example of which there are about 670 in 120 different countries ratified by the United Nations educational scientific and cultural organization. Biosphere reserves have an inner more pristine environment with 2 surrounding zones. The inner core of this biosphere is fully protected from the outside world and managed as close as possible to the historic conditions that existed. Outside of that immediately is a buffer zone with little to no human impact on the functioning of the specific ecosystem there and further beyond that there is a transition zone that is meant to support sustainable agriculture or sometimes Forestry based on the principles in the policy that has been designated for that area but it has very little infrastructure development.

For the purposes of preservation, the transition zone should also be characterized by Low insecticide usc a maximum utilization of all biological controls. They should also be low compaction practices like avoiding the use of heavy machinery and all flora that is planted in the area should be indigenous or natural to that environment. It is only now that research is being carried out on validating these reserves in relation to insert conservation and it continues to grow in importance especially in the 21st century and it can't be a major initiative 4 harmonizing our human activity with our natural surroundings.

Other insect- friendly approaches that go hand-in-hand with agro-ecological perspectives is the idea of land sharing where the landowner or farmer would mix crops an insect- friendly plants for example the planting of coffee in natural forests because the coffee plant needs shade. Another method includes the action of separation of a special scale between conservation areas and production areas which are known as land sparing. There are interconnected conservation corridors however that would develop more ecologically effective and efficient ecological networks like having grasslands next to natural forest patches inside the plantation. These approaches especially when mixed and matched together have proven to be very successful when introducing more diverse ecological networks and by extent actually increasing the size obviously and the protected area will give the system a more resilient standing

These approaches were initially developed in Europe but have become more globally widespread especially with the development of ugly environment schemes where financial compensation is actually provided to farmers who associate themselves with land sharing or land sparing setting aside land without cultivating it. The policies psychology involves maximizing the opportunities for the local indigenous biodiversity to grow while allowing farmers to optimize production. The objective here is not only to conserve and protect the insects and other wildlife but it is to father the natural ecosystem that exists with its various integrated services such as pollination that we have detailed explicitly above and biological control of pests.

Policies like these have been beneficial for insect conservation when practiced efficiently despite the concern of farmers being a drop in production. With increasing pressures on both the global and local scale on eco-friendly farming and forestry approaches the future of insect conservation is bright and we will see some approaches many of them being quite advanced becoming increasingly prevalent in the next coming years.

Another reason why insect conservation policies will become more increasingly successful is that already more than half of the human population live within cities and more rural areas can be terraformed into functioning ecosystems very close to the historic counterpart. There is however another fundamental issue to consider in that respect period the first is that the human intensification on urban landscapes has pigs and a big disconnect between humans and nature. We may find out in the future human intensification in urban areas will be disproportionately valued with nature and the space that it has. Because of this reality, it is unfortunate to say but only the more widespread and generalist insect species will be able to have a strong chance to survive especially with habitat face becoming more and scarcer. To add on that there will also be some major genetic changes that will occur to those insects that would of die out having to adapt to a more strenuous ecosystem.

Insect conservation policies will affect insects in that way especially concerning provisions in all the stages of life and the cycles that they go through. For example, a butterfly will go around feeding on nectar and provide them for the adults while additionally providing sustenance for the larva as well. It will also be interesting to see the massive population dynamics and how they will be maintained because as some species can exist in large urban parks many will require a conservation corridor to promote some sense of genetic diversity

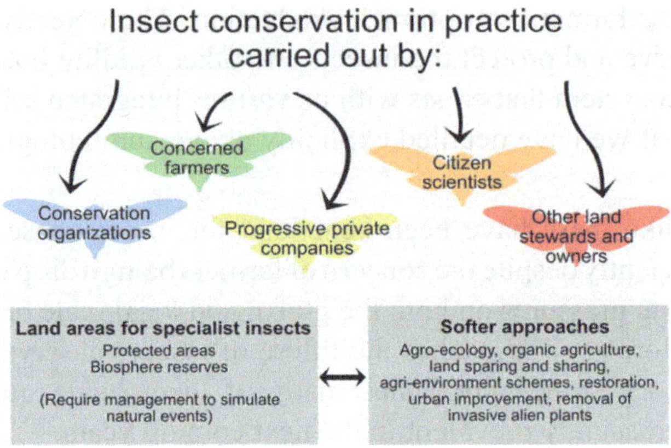

Figure 10.5: Insect conservation is transferred from policy to practice is carried out by different organizational communities. Land cores, like protected areas and the inner zones of reserves, are critical for many insect species. These areas often require some management to recreate natural conditions. Outside of these land preserves, other approaches to landscape design and management can be used as insect conservation action. These approaches include agro-ecology, organic agriculture, the land sparing-land sharing techniques, agro-environment schemes, restoration, urban improvement, and removal of invasive non- indigenous plants.

Source: https://www.intechopen.com/books/insect-science-diversity-conservation-and-nutrition/insect-conservation-for-the-twenty-first-century

Insect conservation policies do not also take into account the temperature increase that is occurring globally and not only in urban environments and it creates yet another challenge. Road kills from the above- ground vehicles: there's also something that is easily overlooked with air transport becoming more and more intensified period on top of that population just as it is an issue for human beings is an issue for insects as well. Another factor is the huge effects of like especially on night-flying insects. These factors are often overlooked but good urban development, planning can be taken into account to consider insects' preservation and protect them from the infamous great insects dying.

In conclusion, the time limit to act is yesterday when it comes to conserving and protecting insects and the necessary insect diversity that has continued to sustain us on this planet. Future generations will look back on this content desperation of the planet and cry out on the infringements we have post up on the delicate web of life that exists. Insects and other biodiversity play a huge and sometimes in the comprehensible room here on earth and even if the ethical based in which we operate is drawing it is improving much too slowly. We also participate in a lot of insect conservation research but again much too slowly and we put in place policies regulations and legislations and again Act too slowly. When the realization that we live much too virtually and too disconnected from nature is well ingrained maybe then the policies on insect protection preservation and Conservation will be physically worked on from the Ground Up in an acceptable fashion.

References

1. Barbour, J. (2016). Sampling Insect Populations for Pest Management . Idaho: University of Idaho.
2. Bergeson, L. L. (2000). FIFRA: Federal Insecticide, Fungicide, and Rodenticide Act. New York: American Bar Association, pp150.
3. Cheesman, O. D., & Key, R. S. (2007). The extinction of experience: a threat to insect conservation. Insect Conservation Biology', (Eds AJ Stewart, TR New, and OT Lewis.) pp, 322-348.
4. Christer Bjorkman & Pekka Niemela. (2015). Climate Change and Insect Pests. London: CABI, pp291.
5. Colla, S. R. (2017). Questioning public perception, conservation policy, and recovery actions for honeybees in North America. Conservation Biology, 31(5), 1202-1204.
6. Collins, N. (2012). The Conservation of Insects and Their Habitats. New York: Academic Press, 33-71.
7. David Kothamasi & E. Toby Kiers. (2009). Emerging Conflicts between Biodiversity Conservation Laws and Scientific Research: the Case of the Czech Entomologists in India. New Delhi: Centre for Environmental Management of Degraded Ecosystems.
8. Didham, R. K., Barbero, F., Collins, C. M., Forister, M. L., Hassall, C., Leather, S. R., ... & Stewart, A. J. . (2020). Spotlight on insects: trends, threats and conservation challenges. .Insect Conservation and Diversity, 13(2), 99-102.
9. Gunther, F. A. (2012). Residues of Pesticides and Other Contaminants in the Total Environment. Berlin: Springer Science & Business Media, 3-45.

10. Hobbs, S. J. (2012). Community Participation In Biodiversity Monitoring. New York: University of New York, pp214.
11. J. F. Campbell, M. A. Mullen, And A. K. Dowdy. (2002). Monitoring Stored-Product Pests in Food Processing Plants with Pheromone Trapping, Contour Mapping, and Mark-Recapture. Grain Marketing and Production Research Center, 95(5): 1089Ð1101.
12. Jack E. Rechcigl & Nancy A. Rechcigl. (2016). Insect Pest Management: Techniques for Environmental Protection. Florida: CRC Press, 3-25.
13. Konvalina, P. (2016). Organic Farming: A Promising Way of Food Production. Norderstedt, Germany: BoD – Books on Demand, 29-83.
14. Kumar, L., Yogi, M. K., & Jagdish, J. (2013). Habitat manipulation for biological control of insect pests: A review. . Research Journal of Agriculture and Forestry Sciences, ISSN, 2320, 6063.
15. Letcher, T. M. (2009). Climate Change: Observed impacts on Planet Earth. New York: Elsevier, pp492.
16. Lewis H. Ziska & Jeffrey S. Dukes. (2014). Invasive Species and Global Climate Change. London: CABI, 1-79.
17. Lugo, E. (2006). Insect Conservation under the Endangered Species Act. UCLA Journal of Environmental Law and Policy, 25(1).
18. Medeiros et al. (2013). The Importance of Insect Monitoring to Conservation Actions in Hawaii. Proceedings of the Hawaiian Entomological Society, 45:149–166.
19. New, T. R. (2012). Insect Conservation: Past, Present and Prospects. Berlin: Springer Science & Business Media, 1-20.
20. New, T. R. (2016). Alien Species and Insect Conservation. Berlin: Springer, 1-99.
21. Reddy, P. P. (2017). Agro-ecological Approaches to Pest Management for Sustainable Agriculture. Berlin: Springer, 1-90.
22. Samways, M. J. (2019). Insect Conservation: A Global Synthesis. London: CABI, 1-85.
23. Saunders, M. E., Janes, J. K., & O'Hanlon, J. C. (2020). Semantics of the insect decline narrative: recommendations for communicating insect conservation to peer and public audiences. Insect Conservation and Diversity, 13(2), 211-213.
24. Scott N. Johnson & T. Hefin Jones. (2017). Global Climate Change and Terrestrial Invertebrates. New Jersey: John Wiley & Sons, 11-71.
25. Soloneski, S. (2014). Pesticides: Toxic Aspects. Norderstedt, Germany:

BoD – Books on Demand, pp240.

26. Timothy T. Marrs & Bryan Ballantyne. (2004). Pesticide Toxicology and International Regulation. Florida: John Wiley & Sons, 1-65.
27. Vonnie D.C. Shields. (2017). Biological Control of Pest and Vector Insects. Norderstedt, Germany: BoD – Books on Demand, 3-71.
28. Wheeler, W. B. (2002). Pesticides in Agriculture and the Environment. Florida: CRC Press, pp360.
29. Yu, S. J. (2011). The Toxicology and Biochemistry of Insecticides. Florida: CRC Press, 1-87.

INDEX

N

O

P

R

Printed in the United States
By Bookmasters